Geodesic and Horocyclic Trajectories

Universitext

For other titles in this series, go to
www.springer.com/series/223

Françoise Dal'Bo

Geodesic and Horocyclic Trajectories

 Springer

Françoise Dal'Bo
IRMAR
Université Rennes 1
Rennes Cedex
France
francoise.dalbo@univ-rennes1.fr

Translation from the French language edition:
Trajectoires géodésiques et horocycliques
by Françoise Dal'Bo
EDP Sciences ISBN 978-2-86883-997-8
Copyright © 2007 EDP Sciences, CNRS Editions, France.
http://www.edpsciences.org/
http://www.cnrseditions.fr/
All Rights Reserved

ISBN 978-0-85729-072-4   e-ISBN 978-0-85729-073-1
DOI 10.1007/978-0-85729-073-1
Springer London Dordrecht Heidelberg New York

British Library Cataloguing in Publication Data
A catalogue record for this book is available from the British Library

Library of Congress Control Number: 2010937998

Mathematics Subject Classification (2010): 37Bxx, 11Jxx, 20H10, 37D40, 11A55

Cover design: deblik

Printed on acid-free paper

Springer is part of Springer Science+Business Media (www.springer.com)

To Dominique,
to Alma and Romance

Preface

In this text, we present an introduction to the topological dynamics of two classical flows associated with surfaces of curvature -1, namely the geodesic and horocycle flows. Since the end of the nineteenth century, many texts have been written on this subject.

Why have we undertaken this project?

In the course of several talks that we have given on this topic, notably during some summer workshops which were organized by the University of Savoie, we have often regretted not being able to recommend a book to those who wanted to find out about this area for themselves. Due to their determination and their enthusiasm for the subject, we decided to go beyond the stage of regret and write up our notes.

In the past thirty years, some very strong connections have been established between dynamical systems and number theory. The intersection of these two fields relies on a change in point of view which, in dimension 2, essentially consists of considering a real number to be a point on the *boundary at infinity* of the Poincaré half-plane and of associating this point to a geodesic, on the modular surface, pointing in its direction (Sect. VII.3). This case study is still the source of inspiration for a large number of specialists. It is sometimes so present in our minds that it is absent from our texts. One of our motivations has been to put this idea back in the spotlight.

Who does this book address?

The reader is expected to have some knowledge of differential geometry and topological dynamics. Our goal has been to produce a text which is readable by a motivated student. Experts in other areas who are interested in this subject have also been considered.

We have attempted to keep the reader from being overwhelmed by proofs which are either too detailed or too succinct by punctuating the text with exercises.

In what spirit was the text written?

This text has been written primarily with the idea of highlighting, in a relatively elementary framework, the existence of gateways between some mathematical fields, and the advantages of using them.

We have chosen not to address the historical aspects of this field and to reserve most of the references until the end of each chapter in the Comments section. Some of our proofs have been borrowed from the literature. In some cases, we have worked to simplify those proofs.

Since the degree of difficulty (or of simplicity) of each chapter is relatively similar, the unfolding of this text may not appear to reach a climax. The applications are the true focus of its progression.

What does the text cover?

We begin with a chapter introducing the geometry of the hyperbolic plane and the dynamics of *Fuchsian groups*, inspired by S. Katok's book "Fuchsian Groups" [41], and A. Katok's and V. Climenhaga's "Lectures on Surfaces" [39]. The action of a Fuchsian group on the Poincaré half plane $\mathbb{H}$ is properly discontinuous. If it is not finite, its orbits accumulate on the boundary at infinity $\mathbb{H}(\infty)$ of $\mathbb{H}$ in a constellation of points called the *limit set* of the group. We focus on several ways in which these points are approximated, which requires us to define *conical* and *parabolic* points, and to introduce the notion of *geometrically finite* groups (Sect. I.4).

In Chap. II, we study some examples of these groups with special attention given to *Schottky groups* and the *modular group*. In each of these cases, we construct a method of encoding their limit sets into sequences and establish a one-to-one correspondence between certain geometric properties of limit points and other combinatorial properties of sequences. For example, the coding introduced for the modular group allows us to interpret the continued fraction expansion of the real numbers in terms of hyperbolic geometry. This coding also allows us to identify a relationship between the *golden ratio* and the length of the shortest compact geodesic on the *modular surface* (Sect. II.4).

In Chap. III, we study the topological dynamics of the geodesic flow $g_{\mathbb{R}}$ on the quotient T^1S of the unitary tangent bundle of $\mathbb{H}$ by a Fuchsian group Γ. The main idea here is to connect the dynamics of this flow to that of the action of Γ on $\mathbb{H}(\infty)$. We show that if Γ is not elementary, the set of periodic elements with respect to $g_{\mathbb{R}}$ is dense in the *non-wandering set* $\Omega_g(T^1S)$ of this flow, and that there exist some trajectories which are dense in $\Omega_g(T^1S)$ (Sects. III.3 and III.4). Also, when Γ is geometrically finite, we construct a compact set which intersects every trajectory in $\Omega_g(T^1S)$ (Sect. III.2).

In Chap. IV, we restrict ourselves to the case in which the Fuchsian group is a Schottky group. Using the coding of its limit set constructed in Chap. II, we develop a symbolic approach which allows us to study the topology of trajectories of the geodesic flow on $\Omega_g(T^1S)$ and to appreciate its complexity. For example, we construct some trajectories in $\Omega_g(T^1S)$ which are neither compact nor dense, and we obtain, in the general case of non-elementary Fuchsian groups, the existence of non-periodic, minimal compact sets which are invariant with respect to the geodesic flow (Sect. IV.3).

Chapter V is devoted to the study of the horocycle flow $h_{\mathbb{R}}$ on T^1S. The method that we use relies on a correspondence between horocycles of $\mathbb{H}$ and non-zero vectors in $\mathbb{R}^2$, modulo $\pm\,\mathrm{Id}$. This vectorial point of view allows us to link the action of $h_\mathbb{R}$ on T^1S to that of a linear group on a vector space, and to determine, for example, the existence of trajectories which are dense in the non-wandering set $\Omega_h(T^1S)$ of this flow (Sects. V.2 and V.3). When the group Γ is geometrically finite, the dynamics of the horocycle flow, unlike that of the geodesic flow, is simple since a trajectory in $\Omega_h(T^1S)$ is either dense or periodic (Sect. V.4). Despite this very different behavior, these two flows are intimately related in the sense that the flow $h_\mathbb{R}$ reflects the collective behavior of asymptotic trajectories of the flow $g_\mathbb{R}$.

The last two chapters are dedicated to some applications of the study of these flows, one in the area of linear actions, the other in that of Diophantine approximations.

In Chap. VI, we focus on the *Lorentz* space $\mathbb{R}^3$ equipped with a bilinear form of signature $(2,1)$. We connect the topology of orbits of a discrete group G of orthogonal transformations of this form to that of the trajectories of the geodesic and horocycle flows on the quotient of the unitary tangent bundle of $\mathbb{H}$ over a Fuchsian group. Translating the results proved about the horocycle flow into this vectorial context, one obtains, for example, a complete description of the orbits of G located in the *light cone*, when this group is of finite type (Sect. VI.3).

In Chap. VII, we translate the Diophantine approximation of a real number by rational ones into the terms of hyperbolic geometry. Relying on the dynamics of the geodesic flow on the modular surface, we rediscover among other things that a real number is *badly approximated* if and only if the coefficients involved in its continued fraction expansion are bounded (Sect. VII.3).

We have chosen to make geometric simplicity a priority and to avoid discussing the metric aspects of these flows. As such, we have limited the scope of some statements and have occasionally hidden some important ideas in these arguments. So that the reader does not come away with the impression that we have said everything there is to say about these subjects, at the end of each chapter we have added some comments in which we recast our treatment of the material into a general Riemannian context and introduce the reader to the vast field of ergodic geometry. We conclude these comments with some open problems as a reminder that this general area has its share of unknown answers and that it has a place in contemporary research.

We are indebted to Claude Sabbah for his attention to the presentation of this text and for his precise re-readings of it. We are equally indebted to Raymond Séroul for creating the figures and to Steven Broad for the translation of the original French text into English. We would also like to thank the referees of the French and English versions for useful comments and suggestions.

Françoise Dal'Bo

Contents

I

Dynamics of Fuchsian groups

This chapter is an introduction to the planar hyperbolic geometry. There are many books which cover it. Our text is inspired by three of them: A. Beardon's "The geometry of discrete groups" [7], A. Katok's and V. Climenhaga's "Lectures on Surfaces" [39], and S. Katok's "Fuchsian groups" [41]. The reader will find in these books the solutions of the exercises suggested in this chapter.

We assume that the reader has some background in complex analysis and differential geometry. For a short introduction to Riemannian geometry, see Appendix B.

Sections 3 and 4 do not include many examples. Readers who prefer to see examples of Fuchsian groups before studying their properties are invited to browse through Chap. II.

1 Introduction to the planar hyperbolic geometry

We follow the conformal approach. Recall that a diffeomorphism ψ between two open subsets U and V of the affine Euclidean plane $\mathbb{R}^2$ is *conformal* if it preserves the oriented angles. More precisely, ψ is conformal if there exists a map f from U to $\mathbb{R}_+^*$ such that for any point x in U and any vectors $\overrightarrow{u}, \overrightarrow{v}$ in the Euclidean plane $\mathbb{R}^2$, equipped with the standard scalar product $\langle \cdot, \cdot \rangle$, we have:

$$\langle T_x\psi(\overrightarrow{u}), T_x\psi(\overrightarrow{v}) \rangle = f(x)\langle \overrightarrow{u}, \overrightarrow{v} \rangle.$$

When $U = V$, we denote by $\mathrm{Conf}(U)$ the group of conformal diffeomorphisms on U. In terms of complex geometry, this group coincides with the group of biholomorphic transformations on U.

Let us describe $\mathrm{Conf}(U)$ when U is the open unit disk $\mathbb{D}$. Observe that the group of Möbius transformations of the form

$$h_{\alpha,\beta}(z) = \frac{\alpha z + \beta}{\overline{\beta} z + \overline{\alpha}},$$

F. Dal'Bo, *Geodesic and Horocyclic Trajectories*, Universitext,
DOI 10.1007/978-0-85729-073-1_1, © Springer-Verlag London Limited 2011

where α and β are complex numbers with $|\alpha|^2 - |\beta|^2 = 1$, is included in $\mathrm{Conf}(\mathbb{D})$. Moreover we have:

Proposition 1.1. *The Möbius transformations of the form $h_{\alpha,\beta}$, where α and β are complex numbers satisfying $|\alpha|^2 - |\beta|^2 = 1$, are the only one conformal diffeomorphisms on $\mathbb{D}$.*

Proof. Let $\psi \in \mathrm{Conf}(\mathbb{D})$. There exists $h_{\alpha,\beta}$ such that the map $\phi = h_{\alpha,\beta}\psi$ fixes the center of $\mathbb{D}$. Applying the classical *Schwartz Lemma* to the holomorphic maps ϕ and ϕ^{-1}, we obtain that $|\phi(z)| \leqslant |z|$ and $|\phi^{-1}(z)| \leqslant |z|$ for any $z \in \mathbb{D}$. It follows that $|\phi(z)| = |z|$. Applying again the same Lemma, we conclude that $\phi = h_{\exp i\theta, 0}$ for some $\theta \in \mathbb{R}$, and hence that ψ is a Möbius transformation of the form $h_{\alpha,\beta}$. $\qquad\square$

Consider now the open half-plane $\mathbb{H} = \{z \in \mathbb{C} \mid \mathrm{Im}\, z > 0\}$. This set is conformal to $\mathbb{D}$ since the map

$$\Psi : \mathbb{H} \longrightarrow \mathbb{D}$$
$$z \longmapsto i\frac{z-i}{z+i}$$

is a holomorphic diffeomorphism. In particular $\mathrm{Conf}(\mathbb{H}) = \Psi^{-1}\mathrm{Conf}(\mathbb{D})\Psi$. Moreover, one checks that for any $h_{\alpha,\beta}$ in $\mathrm{Conf}(\mathbb{D})$, the map $h = \Psi^{-1}h_{\alpha,\beta}\Psi$ is a Möbius transformation of the form:

$$h(z) = \frac{az+b}{cz+d},$$

where a,b,c,d are real numbers satisfying $ad - bc = 1$. We denote by G the group of such real Möbius transformations. We have:

Corollary 1.2. *The group G coincides with $\mathrm{Conf}(\mathbb{H})$.*

Since G contains all the transformations of the form $h(z) = az + b$, with $a > 0$ and $b \in \mathbb{R}$, the group G acts clearly transitively on $\mathbb{H}$:

Property 1.3. *For any z and z' in $\mathbb{H}$, there exists $g \in G$ such that $z' = g(z)$.*

Observe that G does not acts simply transitively on $\mathbb{H}$. In particular the stabilizer of i in G is the group K defined by:

$$K = \left\{ r(z) = \frac{z\cos\theta - \sin\theta}{z\sin\theta + \cos\theta} \;\middle|\; \theta \in \mathbb{R} \right\}.$$

Let $h \in G$, write $h(z) = (az+b)/(cz+d)$, where $ad - bc = 1$. Notice that for any point z in $\mathbb{H}$ and vectors $\overrightarrow{u}$, $\overrightarrow{v}$ in the Euclidean plane $\mathbb{R}^2$, we have (Fig. I.1):

$$\mathrm{Im}\, h(z) = \frac{\mathrm{Im}\, z}{|cz+d|^2} \quad \text{and} \quad \langle T_z h(\overrightarrow{u}), T_z h(\overrightarrow{v})\rangle = \frac{1}{|cz+d|^4}\langle \overrightarrow{u}, \overrightarrow{v}\rangle.$$

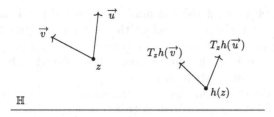

Fig. I.1.

It follows that:

$$\frac{1}{\operatorname{Im} h(z)^2}\langle T_z h(\overrightarrow{u}), T_z h(\overrightarrow{v})\rangle = \frac{1}{\operatorname{Im} z^2}\langle \overrightarrow{u}, \overrightarrow{v}\rangle.$$

We deduce from this formula that the group G acts by isometries on $\mathbb{H}$, if we replace the global Euclidean scalar product $\langle\,,\,\rangle$ by the scalar product g_z depending on each $z \in \mathbb{H}$ and defined on each tangent plane $T_z\mathbb{H}$ by:

$$g_z(\overrightarrow{u}, \overrightarrow{v}) = \frac{1}{\operatorname{Im} z^2}\langle \overrightarrow{u}, \overrightarrow{v}\rangle.$$

The family of $(g_z)_{z\in\mathbb{H}}$ defines a Riemannian metric on $\mathbb{H}$, called the *hyperbolic metric*.

Throughout this text, we consider $\mathbb{H}$ equipped with the metric $(g_z)_{z\in\mathbb{H}}$ and call it the *Poincaré half-plane*.

By construction the angles defined by this metric are the same as the Euclidean one, and hence the group G is included in the group of orientation preserving isometries of $\mathbb{H}$ which we will simply refer to as the group of *positive isometries*.

By recalling some facts from Euclidean geometry, we observe that the hyperbolic metric on $\mathbb{H}$ allows us to define new notions of length and area on each tangent plane $T_z\mathbb{H}$. Namely, if $\overrightarrow{u}$ is in $T_z\mathbb{H}$ its length is $\sqrt{g_z(\overrightarrow{u}, \overrightarrow{u})}$, and the area of a parallelogram is its Euclidean area divided by $\operatorname{Im} z^2$.

These notions give rise to the following global definitions: the *hyperbolic length* of a parametric piecewise-smooth curve $c : [a, b] \to \mathbb{H}$, with $c(t) = x(t) + iy(t)$, is defined by

$$\text{length}(c) = \int_a^b \frac{\sqrt{x'(t)^2 + y'(t)^2}}{y(t)}\,dt,$$

and the *hyperbolic area* of a domain $B \subset \mathbb{H}$ is defined by

$$\mathcal{A}(B) = \iint_B \frac{dx\,dy}{y^2},$$

when this integral exists.

One can check that all these definitions do not depend on a particular parametrization of the curve c and of the domain B. Thus the notion of hyperbolic length is well-defined for piecewise-smooth geometric curves (by *geometric curve*, we mean the image—sometimes also called the trace—of the curve which is a set of points in $\mathbb{H}$.)

Notice that the hyperbolic length of the segment $[ib, a + ib]$ with $b > 0$ is $|a|/b$; likewise, for the segment $[i, ib]$ with $b > 0$, it is $|\ln b|$.

Clearly, the group G preserves all these notions. Namely for any $g \in G$ we have:

$$\text{length}(g(c)) = \text{length}(c) \quad \text{and} \quad \mathcal{A}(g(B)) = \mathcal{A}(B).$$

1.1 Geodesics and distance

Let us now define the analogue in $\mathbb{H}$ of a basic geometric object in the Euclidean affine plane, a straight line. Recall that the Euclidean segment between two points in the plane is the shortest curve between them.

As subgroup of Möbius transformations, the group G acts on the extended complex plane $\mathbb{C} \cup \{\infty\}$, and preserves the family of circles (we regard straight lines in $\mathbb{C}$ as being circles in $\mathbb{C} \cup \{\infty\}$ which pass through ∞). Moreover the circle $\mathbb{R} \cup \{\infty\}$ is globally invariant by G and G preserves the angles. It follows that G preserves the subfamily of vertical half straight lines and half-circles orthogonal to the real axis included in $\mathbb{H}$.

Let z and z' be in $\mathbb{H}$, denote by S the set of piecewise-smooth parametric curves in $\mathbb{H}$ having z and z' as endpoints.

Proposition 1.4. *There exists a unique piecewise-smooth geometric curve C with endpoints z and z' satisfying* $\text{length}(C) = \inf_{c \in S} \text{length}(c)$.

- *If* $\text{Re}(z) = \text{Re}(z')$, *the curve C is the line segment with endpoints z and z'.*
- *Otherwise, consider the half-circle in $\mathbb{H}$ which passes through z and z' and is centered on the real axis. Then C is the arc of this half-circle having endpoints z and z'.*

Proof. We will begin with the case in which $z = is$ and $z' = is'$, where $s > 0$ and $s' > 0$. Let $c : [a, b] \to \mathbb{H}$ be a piecewise-smooth curve with endpoints z and z'. Define $c(t) = x(t) + iy(t)$.

We have:

$$\text{length}(c) = \int_a^b \frac{\sqrt{x'(t)^2 + y'(t)^2}}{y(t)} \, dt \geqslant \left| \int_a^b \frac{y'(t)}{y(t)} \, dt \right|$$

with equality if and only if $x(t) = 0$ for all t in $[a, b]$, and y' does not change sign. Thus $\text{length}(c) \geqslant |\ln(s/s')|$ with equality if and only if $c([a, b])$ is the segment $[is, is']$.

Now let z and z' be any two points. If $\text{Re}(z) = \text{Re}(z')$, using a translation, we deduce from the previous case that the segment $[z, z']$ is the unique curve C in S such that length$(C) = \inf_{c \in S}$ length(c). Otherwise, since G acts transitivity on $\mathbb{H}$, there exists $g \in G$ such that $g(z) = i$. Moreover there exists $k \in K$ such that $kg(z) = i$ and $kg(z')$ is in the positive imaginary axis. It follows from the first case that the shortest curve between $kg(z) = i$ and $kg(z')$ is the segment $[i, kg(z')]$. Since the group G acts on $\mathbb{H}$ by isometries, and preserves the family of vertical half-lines and half-circles centered on the real axis, we obtain that $g^{-1}k^{-1}[i, kg(z')]$ is the shortest curve between z and z', and is included in a half-circle in $\mathbb{H}$ centered on the real axis. $\square$

Definition 1.5. Vertical half-lines and Euclidean half-circles centered on the real axis in $\mathbb{H}$ are called *geodesics* (Fig. I.2).

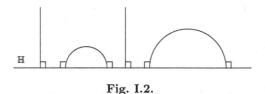

Fig. I.2.

With this characterization of geodesics, we can immediately see that the Euclid's parallel postulate fails in $\mathbb{H}$; given for example the point i and the vertical geodesic $C = \{z \in \mathbb{H} \mid \text{Re}(z) = 2\}$, there are many geodesics passing through i which do not intersect C.

Given two points z and z' in $\mathbb{H}$, the circular arc or line segment with endpoints z and z' contained in the geodesic passing through these two points is called the *hyperbolic segment* and is denoted $[z, z']_h$ (Fig. I.3). This segment is therefore the shortest (in the sense of the hyperbolic metric) piecewise-smooth geometric curve with endpoints z and z'.

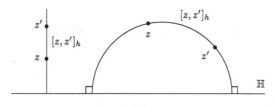

Fig. I.3.

By analogy with the Euclidean affine plane, we define the *hyperbolic distance* between z and z' as follow:

Proposition 1.6. *The function* $d : \mathbb{H} \times \mathbb{H} \to \mathbb{R}^+$ *defined by*

$$d(z, z') = \inf_S \text{length}(c)$$

is a distance function.

Exercise 1.7. Prove Proposition 1.6.

Notice that, by construction, we have for any g in G:

$$d(g(z), g(z')) = d(z, z').$$

For some specific points z and z', the distance $d(z, z')$ is easy to calculate using Proposition 1.4. This is the case when $z = it$ and $z' = it'$ with $t, t' > 0$: one has $d(it, it') = |\ln(t/t')|$. Observe that, when t goes to 0 or to $+\infty$, the points i and it are moving away from each other. On the other hand, $d(it, it + 1)$ is less than $1/t$, meaning that the points it and $it + 1$ are getting closer together as t goes to $+\infty$.

The following exercise suggests a formula relating hyperbolic distance and Euclidean distance. The hyperbolic sine function, denoted by sinh, is defined for real numbers x by

$$\sinh(x) = \frac{e^x - e^{-x}}{2}.$$

Exercise 1.8. Let z and z' be points in $\mathbb{H}$. Show that the following equality holds:

$$\sinh\left(\frac{1}{2}d(z, z')\right) = \frac{|z - z'|}{2(\operatorname{Im} z \operatorname{Im} z')^{1/2}}.$$

(Hint: prove that both sides are invariant under G, and reduce this problem to the case of purely imaginary z and z' [41, Theorem 1.2.6].)

Exercise 1.9. Let z and z' be distinct points in $\mathbb{H}$. Prove that the set of points z'' in $\mathbb{H}$ satisfying $d(z'', z) = d(z'', z')$ is the geodesic passing through the hyperbolic midpoint of $[z, z']_h$, orthogonal to this hyperbolic segment. This set is called the *perpendicular bisector* of $[z, z']_h$.
(Hint: reduce the problem to the case where $\operatorname{Im} z = \operatorname{Im} z$ and use Exercise 1.8.)

We end this subsection with the construction of a distance on the unitary tangent bundle of $\mathbb{H}$ defined by

$$T^1\mathbb{H} = \{(z, \overrightarrow{u}) \mid z \in \mathbb{H},\ \overrightarrow{u} \in T_z\mathbb{H} \text{ and } g_z(\overrightarrow{u}, \overrightarrow{u}) = 1\}.$$

This distance will be useful in Chaps. III and V. Let $(z, \overrightarrow{v})$ be an element of $T^1\mathbb{H}$. Denote by $(v(t))_{t \in \mathbb{R}}$ the unique geodesic through z satisfying $v(0) = z$ and $dv/dt(0) = \overrightarrow{v}$ which is parametrized by (hyperbolic) arclength (i.e., $d(v(t), v(t')) = |t - t'|$) (Fig. I.4). Such a geodesic is sometimes less formally called a "unit speed" geodesic.

Given two elements $(z, \overrightarrow{v})$ and $(z', \overrightarrow{v}')$ of $T^1\mathbb{H}$, we introduce the function $f : \mathbb{R} \to \mathbb{R}^+$ defined by $f(t) = d(v(t), v'(t))e^{-|t|}$. Clearly f satisfies the following inequality

$$f(t) \leqslant (2|t| + d(v(0), v'(0)))e^{-|t|}.$$

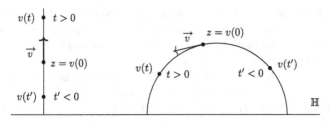

Fig. I.4.

This inequality implies that this function is integrable.

Define

$$D((z, \overrightarrow{u}), (z', \overrightarrow{v}')) = \int_{-\infty}^{+\infty} e^{-|t|} d(v(t), v'(t)) \, dt.$$

Proposition 1.10. *The function* $D : T^1\mathbb{H} \times T^1\mathbb{H} \to \mathbb{R}^+$ *is a distance function that is G-invariant.*

Exercise 1.11. Prove Proposition 1.10.

1.2 Compactification of $\mathbb{H}$

The topology induced on $\mathbb{H}$ by d is the same as the one induced by Euclidean distance. In this topology, $\mathbb{H}$ is not compact. We compactify it by taking its closure in the extended complex $\mathbb{C} \cup \{\infty\}$. The set $\mathbb{H}(\infty) = \mathbb{R} \cup \{\infty\}$ is called the *boundary at infinity* of $\mathbb{H}$. The restriction to $\mathbb{H}$ of the topology on $\mathbb{H} \cup \mathbb{H}(\infty)$ retains the topology induced by d. More precisely, an open set of $\mathbb{H} \cup \mathbb{H}(\infty)$ is either an open set of $\mathbb{H} \cup \mathbb{R}$ (relative to the topology induced by the Euclidean distance on $\mathbb{R}^2$) or the union of the point ∞ and the complement of a compact set in $\mathbb{H} \cup \mathbb{R}$.

Exercise 1.12. Prove that the map

$$\begin{aligned} \Psi : \mathbb{H} &\longrightarrow \mathbb{D} \\ z &\longmapsto i\,\frac{z-i}{z+i} \end{aligned}$$

extends to a homeomorphism between $\mathbb{H} \cup \mathbb{H}(\infty)$ and the closed unit disk of the plane.

Notation. Given a subset A of $\mathbb{H} \cup \mathbb{H}(\infty)$, let $\overset{\circ}{A}$ denote its interior and $\overline{A}$ its closure.

Definition 1.13. The *boundary at infinity of* A is the set denoted by $A(\infty)$ defined by

$$A(\infty) = \overline{A} \cap \mathbb{H}(\infty).$$

The boundary at infinity of a geodesic is a set containing two elements that are called the *endpoints* of the geodesic. Notice that a geodesic is uniquely determined by its endpoints. Also a geodesic is a vertical half-line if and only if one of its endpoints is the point ∞.

Let x^- and x^+ be two distinct points of $\mathbb{H}(\infty)$. Denote by $(x^- x^+)$ the geodesic having endpoints x^-, x^+, oriented from x^- to x^+. If z belongs to $\mathbb{H}$, the geodesic ray originating at z and ending at x^+ is denoted by $[z, x^+)$ (Fig. I.5).

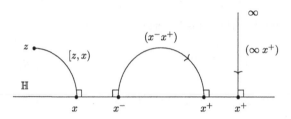

Fig. I.5.

Since G is a subgroup of Möbius transformations, G acts on $\mathbb{C} \cup \{\infty\}$ by homeomorphisms and hence on $\mathbb{H}(\infty)$. More precisely, for $g(z) = (az + b)/(cz + d)$, and x in $\mathbb{R}$ we have:

- if $c = 0$, then $g(\infty) = \infty$ and $g(x) = (ax + b)/d$,
- if $c \neq 0$, then $g(\infty) = a/c$, $g(-d/c) = \infty$ and if $x \neq -d/c$, then $g(x) = (ax + b)/(cx + d)$.

The following exercise is a projective approach of the action of G on $\mathbb{H}(\infty)$. Let $\mathbb{RP}^1$ be the real projective line. For $\overrightarrow{u} \neq \overrightarrow{o}$ in $\mathbb{R}^2$, denote $\mathbb{R}^* \overrightarrow{u}$ its equivalence class in $\mathbb{RP}^1$ (i.e., the non-zero real multiples of $\overrightarrow{u}$).

Exercise 1.14. Prove that the map $\Phi : \mathbb{H}(\infty) \to \mathbb{RP}^1$ defined by

$$\Phi(x) = \begin{cases} \mathbb{R}^* \begin{pmatrix} x \\ 1 \end{pmatrix} & \text{if } x \in \mathbb{R}, \\ \mathbb{R}^* \begin{pmatrix} 1 \\ 0 \end{pmatrix} & \text{if } x = \infty, \end{cases}$$

is a homeomorphism, and that the action of G on $\mathbb{H}(\infty)$ is conjugate to the action of the group $\{\pm \mathrm{Id}\} \backslash \mathrm{SL}(2, \mathbb{R}) = \mathrm{PSL}(2, \mathbb{R})$ on $\mathbb{RP}^1$.

1.3 Hyperbolic triangles and circles

Each geodesic of $\mathbb{H}$ separates it into two connected components. Each of these components is called a *half-plane*. By definition, a *hyperbolic triangle* T is the intersection of three closed half-planes whose hyperbolic area is finite and non-zero. The following proposition is technical but classical, for its proof see [7, Chap. 7], [41, Theorem 1.4.2], [39, Lecture 30].

Proposition 1.15. *Let T be a hyperbolic triangle, its boundary at infinity $T(\infty)$ contains at most three points. Furthermore, let $\mathcal{F}(T) = T - \overset{\circ}{T}$ (Fig. I.6), we have*

- *if $T(\infty) = \{x_1, x_2, x_3\}$, then*

$$\mathcal{F}(T) = (x_1 x_2) \cup (x_2 x_3) \cup (x_3 x_1) \quad and \quad \mathcal{A}(T) = \pi,$$

- *if $T(\infty) = \{x_1, x_2\}$, then there is $z \in \mathbb{H}$ such that*

$$\mathcal{F}(T) = (x_1, z] \cup [z, x_2) \cup (x_2, x_1) \quad and \quad \mathcal{A}(T) = \pi - \alpha,$$

  where α is the angle at z,
- *if $T(\infty) = \{x\}$, then there are z_1 and z_2 in $\mathbb{H}$ such that*

$$\mathcal{F}(T) = (x, z_1] \cup [z_1, z_2]_h \cup [z_2, x) \quad and \quad \mathcal{A}(T) = \pi - (\alpha_1 + \alpha_2),$$

  where α_i is the angle at z_i, for $i = 1, 2$,
- *if $T(\infty) = \varnothing$, then there are z_1, z_2 and z_3 in $\mathbb{H}$ such that*

$$\mathcal{F}(T) = [z_1, z_2]_h \cup [z_2, z_3]_h \cup [z_3, z_1]_h \quad and \quad \mathcal{A}(T) = \pi - (\alpha_1 + \alpha_2 + \alpha_3),$$

  where α_i is the angle at z_i, for $i = 1, 2, 3$.

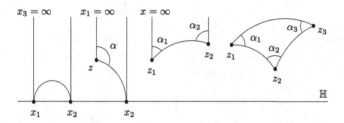

Fig. I.6.

Notice that the previous Proposition shows in particular that, contrary to the Euclidean situation, a hyperbolic triangle is not necessary a compact subset of $\mathbb{H}$, and that the sum of the angles of a compact hyperbolic triangle included in $\mathbb{H}$ is strictly lesser than π.

The *hyperbolic circle* (resp. *hyperbolic disk*) of radius $r > 0$ centered at $z \in \mathbb{H}$ is the set of all z' in $\mathbb{H}$ such that $d(z, z') = r$ (resp. $d(z, z') \leqslant r$). Such sets are the same as the Euclidean one.

Exercise 1.16. Prove that the hyperbolic circle of radius $r > 0$ centered at $z = a + ib$ is the Euclidean circle having the segment $[a + ibe^r, a + ibe^{-r}]$ as a diameter.

Exercise 1.17. Prove that the hyperbolic circumference of any hyperbolic circle of radius $r > 0$ is $2\pi \sinh r$, and that the hyperbolic area of any hyperbolic disk of radius $r > 0$ is $4\pi \sinh^2 r/2$.

Let $K(z)$ denote the *Gauss curvature* at a point z of $\mathbb{H}$. By definition, $K(z)$ measures the difference between the circumference of a Euclidean circle of radius r centered at z, and the hyperbolic circumference $c(r)$ of the hyperbolic circle of the same center and radius, for small r. More precisely, the following formula holds [9, 10.5.1.3], [39, Lecture 32]:

$$K(z) = 3 \lim_{r \to 0} \frac{2\pi r - c(r)}{\pi r^3}.$$

Since $c(r) = 2\pi \sinh r$ (Exercise 1.17), for any $z \in \mathbb{H}$ we have $K(z) = -1$.

1.4 Horocycles and Busemann cocycles

The family of the extended horizontal lines (i.e., with $\{\infty\}$) and of circles tangent at the real line is another family of curves in $\mathbb{H} \cup \mathbb{H}(\infty)$ invariant by G, since this group preserves the family of circles in $\mathbb{C} \cup \{\infty\}$, and $G(\mathbb{H}(\infty)) = \mathbb{H}(\infty)$. There are different approaches to these curves.

One is related to the geodesics of $\mathbb{H}$. Clearly, a horizontal line is orthogonal to the pencil of all vertical geodesics. Replacing this line by a circle tangent at the real line to some point $x \in \mathbb{R}$, and using a transformation $g \in G$ such that $g(\infty) = x$, we obtain that this circle (without x), is orthogonal to the pencil of all geodesics $(x^- x^+)$, with $x^+ = x$.

Such curves can be also viewed as *limit circles*. Namely, using Exercise 1.16, one checks that an extended horizontal line is the limit in $\mathbb{H} \cup \mathbb{H}(\infty)$ of hyperbolic circles passing through a fixed point z in $\mathbb{H}$, with center converging to ∞ along the geodesic ray $[z, \infty)$. The same property holds for a circle tangent at the real line, replacing the point ∞ by the point of tangency. For this reason, an horizontal line or a circle tangent at the real line (without its point of tangency) is usually called an *horocycle*, and its boundary at infinity, its *center*.

We give now a metric approach to these curves, which we will use in the next Chapters. The idea is to sit at a point x on $\mathbb{H}(\infty)$ and observe the points of $\mathbb{H}$ from x. To do this, we will associate to each pair of points z and z' in $\mathbb{H}$, an algebraic quantity reflecting the relative position of these two points, as seen from x.

Theorem 1.18. *Let $(r(t))_{t \geqslant 0}$ be a geodesic ray with endpoint x, parametrized by arclength. For any z and z' in $\mathbb{H}$, the function $f(t) = d(z, r(t)) - d(z', r(t))$ has a limit when t goes to $+\infty$. This limit, called the Busemann cocycle centered at x, calculated at z, z', does not depend on the origin $r(0)$ of the geodesic ray. It is denoted by $B_x(z, z')$. By construction, the function $B_x(z, .)$ is constant along each horocycle centered at x.*

Proof. We want to show that f has a limit at $+\infty$. Let us begin with the case in which $x = \infty$, $z = ib$ and $(r(t))_{t \geqslant 0}$ is the geodesic ray $[z, \infty)$. Let $z' = a' + ib'$ and $s(t) = a' + ibe^t$ (Fig. I.7). For large t, one has

$$d(s(t), z') = \ln(b/b') + t,$$

thus

$$f(t) = d(s(t), z') - d(z', r(t)) + \ln(b'/b).$$

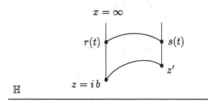

Fig. I.7.

In addition, since $d(s(t), r(t))$ is lesser than the hyperbolic length of the Euclidean segment $[s(t), r(t)]$, we have $d(s(t), r(t)) \leqslant |a'|/be^t$, thus $\lim_{t \to +\infty} f(t) = \ln(b'/b)$.

Replace now $(r(t))_{t \geqslant 0}$ by a geodesic ray $(r''(t))_{t \geqslant 0} = [z'', \infty)$ with $z'' = a'' + ib''$. Using the previous case, we have

$$\lim_{t \to +\infty} d(z'', r''(t)) - d(z, r''(t)) = \ln(b/b'')$$

and

$$\lim_{t \to +\infty} d(z'', r''(t)) - d(z', r''(t)) = \ln(b'/b'').$$

It follows that

$$\lim_{t \to +\infty} d(z, r''(t)) - d(z', r''(t)) = \lim_{t \to +\infty} d(z, r(t)) - d(z', r(t)).$$

Moreover, notice that for z fixed, the limit of $f(t) = d(z, r(t)) - d(z', r(t))$ only depends on Im z'. It follows that the function $B_\infty(z, .)$ is constant along each horizontal line. Furthermore, if $\mathrm{Re}(z') = 0$ then $B_\infty(z, z') = d(z, z')$ for $b' \geqslant b$, and $B_\infty(z, z') = -d(z, z')$ for $b' < b$.

A translation reduces the case in which $x = \infty$ and z is arbitrary to the original case (Fig. I.7).

If x is not ∞, then the Möbius transformation $h \in G$ defined by

$$h(z) = \frac{xz - x^2 - 1}{z - x}$$

similarly recovers the original case. Thus one can conclude that $f(t)$ has a limit at $+\infty$ which does not depend on the origin of the geodesic ray $(r(t))_{t \geqslant 0}$, and that the function $B_x(z, .)$ is constant along each horocycle centered at x (Fig. I.8). $\qquad \square$

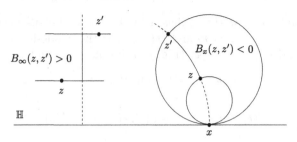

Fig. I.8.

Property 1.19. *Let g be in G, x in $\mathbb{H}(\infty)$, and z, z', z'' in $\mathbb{H}$. One has:*

(i) $B_{g(x)}(g(z), g(z')) = B_x(z, z')$;

(ii) $B_x(z, z') = B_x(z, z'') + B_x(z'', z')$;

(iii) $-d(z, z') \leqslant B_x(z, z') \leqslant d(z, z')$;

(iv) $B_x(z, z') = d(z, z')$ *(resp. $-d(z, z')$) if and only if z' belongs to the ray $[z, x)$ (resp. $z \in [z', x)$).*

Exercise 1.20. Prove Property 1.19.
(Hint: for (ii), (iii), (iv) reduce the problem to the case in which $x = \infty$.)

Notation. For any $t > 0$, the horocycle (resp. *horodisk*) centered at x defined by $\{z \in \mathbb{H} \mid B_x(i, z) = \ln t\}$ (resp. $\{z \in \mathbb{H} \mid B_x(i, z) \geqslant \ln t\}$) is denoted $H_t(x)$ (resp. $H_t^+(x)$) (Fig. I.9).

If $x = \infty$, then $H_t(\infty)$ is the horizontal line defined by $\operatorname{Im} z = t$, and $H_t^+(\infty)$ is the closed Euclidean half-plane in $\mathbb{H}$ bounded by this line. Otherwise, consider g in G satisfying $g(\infty) = x$. From Property 1.19(i) and (ii), we have
$$H_t(x) = g(H_{t'}(\infty)) \quad \text{with } t' = te^{-B_x(i, g(i))}.$$

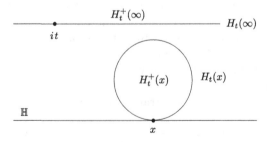

Fig. I.9.

1.5 The Poincaré disk

Let us come back to the open unit disk $\mathbb{D} = \{z \in \mathbb{C} \mid |z| < 1\}$. Remember that this set is the image of $\mathbb{H}$ by the Möbius transformation $\Psi(z) = i(z - i)/(z + i)$. This transformation allows us to transport on $\mathbb{D}$ all the hyperbolic notions defined on $\mathbb{H}$ in the previous sections.

More precisely, we consider the metric on $\mathbb{D}$, already denoted $(g_z)_{z \in \mathbb{D}}$, defined for any $\overrightarrow{u}$ and $\overrightarrow{v}$ in the tangent plane $T_z \mathbb{D}$ by:

$$g_z(\overrightarrow{u}, \overrightarrow{v}) = \frac{1}{|\operatorname{Im} \Psi^{-1}(z)|^2} \langle T_z \Psi^{-1}(\overrightarrow{u}), T_z \Psi^{-1}(\overrightarrow{v}) \rangle.$$

We have:

$$g_z(\overrightarrow{u}, \overrightarrow{v}) = \left(\frac{2}{1 - |z|^2} \right)^2 \langle \overrightarrow{u}, \overrightarrow{v} \rangle.$$

The disk $\mathbb{D}$ equipped with this metric is called the *Poincaré disk*. By construction it is isometric to the Poincaré half-plane. As we seen in Sect. 1.1, the group $\Psi G \Psi^{-1}$ is the group of Möbius transformations of the form $(az + b)/(\bar{b}z + \bar{a})$, with complex coefficients a and b such that $|a|^2 - |b|^2 = 1$. Clearly this group acts by isometries on the Poincaré disk.

We will retain all the notations introduced for $\mathbb{H}$ in our discussion of $\mathbb{D}$. As such, $\Psi G \Psi^{-1}$ is already denoted G, and d represents the distance induced by the metric $(g_z)_{z \in \mathbb{D}}$ on $\mathbb{D}$.

One of the advantages of this model is that its compactification corresponds to the Euclidean one: the boundary at infinity of $\mathbb{D}$ is the unit circle $\mathbb{D}(\infty) = \{z \in \mathbb{C} \mid |z| = 1\}$.

One easy checks that geodesics are circular arcs orthogonal to $\mathbb{D}(\infty)$ or Euclidean diameters of $\mathbb{D}$. Horocycles are circles contained in $\mathbb{D}$ which are tangent to the unit circle (Fig. I.10).

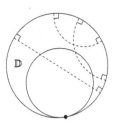

Fig. I.10.

Another advantage of this model is that the Euclidean rotation around the origin is an isometry. It implies for example that a hyperbolic circle of $\mathbb{D}$ with center the origin is an Euclidean circle with the same center (but different radius!).

The Poincaré disk can be also very useful to study properties of hyperbolic triangles, since using a isometry we can reduce our attention to triangles having one vertex at the origin. For such triangles, two of these sides are Euclidean segments and the third one is a part of an Euclidean circle, which is convex in the Euclidean sense. Following this way, it is clear for example that hyperbolic triangles have angles whose sum is less than π, as we already know (see [39, Lecture 30] for another applications).

From now, we will switch back and forth between these two models depending on the type of symmetry for which a particular problem calls.

2 Positive isometries and Fuchsian groups

In this section, we will be interested in the dynamics of the elements of G acting on $\mathbb{H}$. We have already shown that G acts by positive isometries on $\mathbb{H}$.

2.1 Decompositions of the group G

We first introduce the following three subgroups of G:

$$K = \left\{ r(z) = \frac{z \cos\theta - \sin\theta}{z \sin\theta + \cos\theta} \ \middle|\ \theta \in \mathbb{R} \right\},$$
$$A = \{h(z) = az \mid a > 0\},$$
$$N = \{t(z) = z + b \mid b \in \mathbb{R}\}.$$

Recall that K is the stabilizer of i in G. The groups A and N can be also characterized by their fixed points in $\overline{\mathbb{H}} = \mathbb{H} \cup \mathbb{H}(\infty)$.

Exercise 2.1. Prove that an element of G is in A if and only if it fixes 0 and ∞, and that a non-identity element of G is in N if and only if the point ∞ is its only fixed point in $\overline{\mathbb{H}}$.

Observe that elements of A leave the geodesic (0∞) invariant (Fig. I.11). Also, elements of N preserve each horocycle centered at the point ∞ (Fig. I.12).

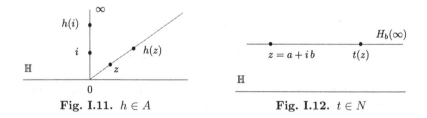

Fig. I.11. $h \in A$ Fig. I.12. $t \in N$

Proposition 2.2. *The group G is the group of positive isometries of $\mathbb{H}$.*

Proof. It is enough to prove that a positive isometry f of $\mathbb{H}$ is in G. Since the action of G on $\mathbb{H}$ is transitive and isometric, after composing f with an element of G, one may assume that f fixes i. Choose $r \in K$ such that $rf(0) = 0$. Then rf maps the geodesic from 0 through i (i.e., the half imaginary axis) to itself, so that $rf(\infty) = \infty$. Thus rf is in A and hence $rf(z) = az$, for some $a > 0$. Since $rf(i) = i$, we have $a = 1$. It follows that f is in G. □

Recall that the unit tangent bundle of $\mathbb{H}$ is defined by

$$T^1\mathbb{H} = \{(z, \overrightarrow{u}) \mid z \in \mathbb{H},\ \overrightarrow{u} \in T_z\mathbb{H} \text{ and } g_z(\overrightarrow{u}, \overrightarrow{u}) = 1\}.$$

For $g(z) = (az + b)/(cz + d)$, where a, b, c, d are real numbers satisfying $ad - bc = 1$, we have

$$g(z, \overrightarrow{u}) = \left(\frac{az + b}{cz + d}, \frac{\overrightarrow{u}}{(cz + d)^2} \right),$$

where multiplying $\overrightarrow{u}$ by a non-zero complex number means taking the image of $\overrightarrow{u}$ by the linear transformation represented by this complex number.

Notice that for $r(z) = (z \cos\theta - \sin\theta)/(z \sin\theta + \cos\theta)$, $h(z) = az$ with $a > 0$, and $t(z) = z + b$ with $b \in \mathbb{R}$, we have

$$r(i, \overrightarrow{u}) = (i, \exp -2i\theta\,\overrightarrow{u}), \quad h(z, \overrightarrow{u}) = (az, a\overrightarrow{u}), \quad t(z, \overrightarrow{u}) = (z + b, \overrightarrow{u}).$$

Property 2.3. *For any $(z, \overrightarrow{v})$ and $(z', \overrightarrow{v}')$ in $T^1\mathbb{H}$, there exists an unique g in G such that $g(z, \overrightarrow{v}) = (z', \overrightarrow{v}')$ (i.e., the action of G on $T^1\mathbb{H}$ is simply transitive).*

Proof. Since the action of G on $\mathbb{H}$ is transitive (Property 1.3), it is enough to prove that for any $\overrightarrow{v}$ and $\overrightarrow{v}'$ in $T_i{}^1\mathbb{H}$, there exists only one $g \in G$ such that $\overrightarrow{v}' = T_i g(\overrightarrow{v})$. Consider the real θ such that -2θ is the measure of the oriented angle between $\overrightarrow{v}$ and $\overrightarrow{v}'$, and set $r(z) = (z \cos\theta - \sin\theta)/(z \sin\theta + \cos\theta)$. We have $r(i, \overrightarrow{v}) = (i, \overrightarrow{v}')$. Now suppose that some g in G fixes $(i, \overrightarrow{v})$, then g is in K and hence $g(z) = (z \cos\theta' - \sin\theta')/(z \sin\theta' + \cos\theta')$, for some real θ'. Since $\exp -2i\theta' \overrightarrow{v} = \overrightarrow{v}$, the real θ' is a multiple of π and hence $g = id$. □

The following proposition gives two decompositions of G along the groups K, A, N. The *Cartan's decomposition* will be used in the Chap. VI (Exercise VI.1.5).

Proposition 2.4. *Let $g \neq id$ in G.*

- *Iwasawa's decomposition: there exists an unique triple (n, a, k) in $N \times A \times K$ such that $g = nak$.*
- *Cartan's decomposition: there exists a (non-unique) triple (k, a, k') in $K \times A \times K$ such that $g = kak'$.*

Proof.
Iwasawa's decomposition: Fix $\vec{u}$ in $T_i\mathbb{H}$. Let g in G, set $g(i, \vec{u}) = (x+iy, \vec{v})$ and consider the transformations: $a(z) = yz$, $n(z) = z + x$ and $k(z) = z\cos\theta - \sin\theta / z\sin\theta + \cos\theta$, where -2θ is the oriented angle between $\vec{u}$ and $\vec{v}$ (Fig. I.13). By construction $nak(i, \vec{u}) = (x + iy, \vec{v})$, and hence $g = nak$

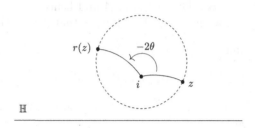

Fig. I.13. $r \in K$

since the action of G on $T^1\mathbb{H}$ is simply transitive.

Suppose now that $nak = n'a'k'$, for $n' \in N$, $a' \in A$ and $k' \in K$. This implies $a'^{-1}n'^{-1}na = k'k^{-1}$ and hence $k = k'$ since $k'k^{-1}(\infty) = \infty$. It follows that $n'^{-1}n = a'a^{-1}$. Using the fact that $N \cap A = \{\text{Id}\}$, we obtain $a = a'$ and $n = n'$.

Cartan's decomposition: Take $g \in G$ and consider the transformation $a \in A$ defined by $a(z) = ze^{d(i,g(i))}$. This map sends i into the hyperbolic circle centered at i passing through $g(i)$. The action of the group K on this circle is transitive, so there is some k in K such that $ka(i) = g(i)$. The isometry $g^{-1}ka$ fixes i and is positive, hence there is a k' in K such that $g = kak'$.

Notice that this decomposition is not unique since for example for k in K, we have $k = k' \, \text{Id} \, k'^{-1}k$ for any k' in K. □

2.2 The dynamics of positive isometries

Now we turn to the task of classifying the positive isometries of $\mathbb{H}$ and understanding what they look like geometrically. We begin by considering g in G and look for fixed points in $\overline{\mathbb{H}} = \mathbb{H} \cup \mathbb{H}(\infty)$.

Write $g(z) = (az + b)/(cz + d)$, where a, b, c, d are real numbers with $ad - bc = 1$. Suppose $g \neq \text{Id}$. If $c \neq 0$, then clearly $z \in \overline{\mathbb{H}}$ is fixed by g if and only if

$$z = \frac{a - d \pm \sqrt{(a + d)^2 - 4}}{2c}.$$

If $c = 0$, then g belongs to A or N. Let us introduce the absolute value of the *trace* of g defined by

$$|\text{tr}(g)| = |a + d|.$$

We have:

Property 2.5. *Let g in $G - \{\mathrm{Id}\}$.*

- *If $|\mathrm{tr}(g)| > 2$, then g fixes exactly two points of $\overline{\mathbb{H}}$, both of which are in $\mathbb{H}(\infty)$, and g is conjugate in G to an element of A.*
- *If $|\mathrm{tr}(g)| < 2$, then g fixes exactly one point of $\overline{\mathbb{H}}$ which is in $\mathbb{H}$, and g is conjugate in G to an element of K.*
- *If $|\mathrm{tr}(g)| = 2$, then g fixes exactly one point of $\overline{\mathbb{H}}$ which is in $\mathbb{H}(\infty)$, and g is conjugate in G to an element of N.*

Exercise 2.6. Prove Property 2.5.

When $|\mathrm{tr}(g)| > 2$, then g is said to be *hyperbolic*. Such an isometry preserves the geodesic having its endpoints at the two fixed points. This geodesic is called the *axis of translation* of g or more simply the *axis* of g. Each hyperbolic g acts on its axis by translation. For any point z in this axis, the sequences $(g^n(z))_{n\geqslant 1}$ and $(g^{-n}(z))_{n\geqslant 1}$ converge to the fixed points of g. The limit of the sequence $(g^n(z))_{n\geqslant 1}$ is its *attractive* fixed point, and the limit of $(g^{-n}(z))_{n\geqslant 1}$ is its *repulsive* fixed point. The attractive (resp. repulsive) fixed point is denoted g^+ (resp. g^-) (Fig. I.14).

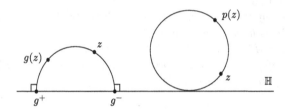

Fig. I.14.

When $|\mathrm{tr}(g)| < 2$, then g is said to be *elliptic*. This transformation fixes an unique point $z \in \mathbb{H}$. Clearly, the image by g of any geodesic passing through z is a geodesic passing through z. Moreover the angle between such a geodesic and its image does not depend on the geodesic. Thus g is analogous to what we term rotation in the Euclidean context.

When $|\mathrm{tr}(g)| = 2$, then g is said to be *parabolic*. Such a g is actually conjugate to a translation $t(z) = z + b$, thus g preserves each horocycle centered at its fixed point, on which it acts by translation (Fig. I.14, $g = p$).

Let us give another approach to these classes of isometries. Let g in $G - \{\mathrm{Id}\}$ and $z_0 \in \mathbb{H}$ which is not fixed by g, recall (Exercise 1.9) that the perpendicular bisector of the hyperbolic segment $[z_0, g(z_0)]_h$ defined by $M_{z_0}(g) = \{z \in \mathbb{H} \mid d(z, z_0) = d(z, g(z_0))\}$, is the geodesic orthogonal to the segment $[z_0, g(z_0)]_h$, passing through its middle.

This geodesic separates $\mathbb{H}$ into two connected components, we denote $D_{z_0}(g)$ the closed half-plane in $\mathbb{H}$ bounded by $M_{z_0}(g)$ containing $g(z_0)$ (Fig. I.15).

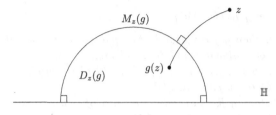

Fig. I.15.

Clearly we have

$$g(M_{z_0}(g^{-1})) = M_{z_0}(g) \quad \text{and} \quad g(\mathring{D}_{z_0}(g^{-1})) = \mathbb{H} - D_{z_0}(g).$$

The type of the transformation g can be characterized by relative position of $M_{z_0}(g)$ and $M_{z_0}(g^{-1})$.

Property 2.7. *Let g in $G - \{\mathrm{Id}\}$ and $z_0 \in \mathbb{H}$ which is not fixed by g.*

- *The geodesics $M_{z_0}(g)$ and $M_{z_0}(g^{-1})$ intersect in $\mathbb{H}$ if and only if g is elliptic (Fig. I.16).*
- *The closures of the geodesics $M_{z_0}(g)$ and $M_{z_0}(g^{-1})$ are disjoint in $\overline{\mathbb{H}}$ if and only if g is hyperbolic (Fig. I.17).*
- *The geodesics $M_{z_0}(g)$ and $M_{z_0}(g^{-1})$ have exactly one endpoint in common if and only if g is parabolic. Furthermore, the common endpoint is the (only) fixed point of g (Fig. I.18).*

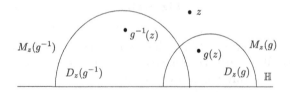

Fig. I.16. g elliptic

Exercise 1. Prove Property 2.7.
(Hint: check that for any $g' \in G$, we have $M_{z_0}(g'gg'^{-1}) = g'(M_{g'^{-1}(z_0)}(g))$, and recover the case in which g is contained in K, A or N.)

The classification of the positive isometries of $\mathbb{H}$ in terms of hyperbolic, elliptic and parabolic transformations can be also obtained using the notion of *displacement* $\ell(g)$ of an isometry g in G defined by

$$\ell(g) = \inf_{z \in \mathbb{H}} d(z, g(z)).$$

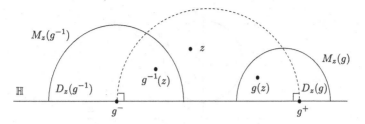

Fig. I.17. g hyperbolic

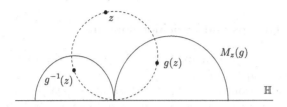

Fig. I.18. g parabolic

Property 2.8. *Let g be in $G - \{\mathrm{Id}\}$.*

- *g is elliptic if and only if $\ell(g) = 0$ and this lower bound is attained.*
- *g is hyperbolic if and only if $\ell(g) > 0$, and $d(z, g(z)) = \ell(g)$ if and only if z is in the axis of g.*
- *g is parabolic if and only if $\ell(g) = 0$ and this lower bound is not attained.*

Proof. Since for any $g' \in G$, we have $\ell(g'gg'^{-1}) = \ell(g)$, we can suppose that g is in K, A or N.

If g is in K, then it fixes the point i and thus $\ell(g) = 0$.

If g is in A, write $g(z) = \lambda z$ with $\lambda > 1$. Since positive dilations and translations are isometries, one has

$$\ell(g) = \inf_{x \in \mathbb{R}} d(i, x + i\lambda).$$

Consider the hyperbolic circle centered at i, passing through $x + i\lambda$ (Fig. I.19). Recall (Exercise 1.16) that this circle is the Euclidean circle having the segment $[ie^r, ie^{-r}]$ as its diameter, where $r = d(i, x + i\lambda)$. It intersects the horizontal line described by the equation $\mathrm{Im}\, z = \lambda$. Thus $\lambda \leqslant e^r$ with equality if and only if $x = 0$. As a result, $d(i, x + i\lambda) \geqslant \ln \lambda$ with equality if and only if $x = 0$. One obtains that $\ell(g) > 0$ and $d(z, g(z)) = \ell(g)$ if and only if z is in the axis of g.

Finally, if g is in N write $g(z) = z + b$. Since the distance $d(z, g(z))$ is smaller than the hyperbolic length of the Euclidean segment $[z, z + b]$, we have $d(z, g(z))) \leqslant |b|/\mathrm{Im}\, z$ and hence $\ell(g) = 0$. Moreover, since g does not fix any point in $\mathbb{H}$, this lower bound is never attained.

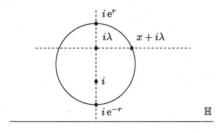

Fig. I.19.

2.3 Fuchsian Groups and Dirichlet domains

Now that we have studied individual positive isometries of $\mathbb{H}$, we will turn our focus to subgroups Γ of G.

Since our motivation is to obtain topologically regular surfaces in the quotient of $\mathbb{H}$ by Γ, we will restrict our interest to so-called *Fuchsian groups*, which are discrete subgroups of G with respect to the topology on G induced by the Euclidean distance on $\mathbb{R}^4$.

We say that the action of a subgroup Γ of G on $\mathbb{H}$ is *properly discontinuous* if for all compact subsets K of $\mathbb{H}$, only finitely many elements γ of Γ satisfy $\gamma K \cap K \neq \varnothing$.

Property 2.9. *The action of a Fuchsian group Γ on $\mathbb{H}$ is properly discontinuous.*

Proof. Let K be a compact subset of $\mathbb{H}$ and denote by K^1 the compact subset of $T^1\mathbb{H}$ composed of elements of the form $(z, \overrightarrow{u})$ where $z \in K$, and $\overrightarrow{u} \in T_z^{\,1}\mathbb{H}$ satisfies $g_z(\overrightarrow{u}, \overrightarrow{u}) = 1$. Fix an element $(z', \overrightarrow{v}')$ in $T^1\mathbb{H}$. The action of G on $T^1\mathbb{H}$ is clearly continuous. Moreover it is simply transitive (Property 2.3), hence there is some compact subset C of G satisfying $C((z', \overrightarrow{v}')) = K^1$. Consider an element γ of Γ. If $\gamma K \cap K \neq \varnothing$, then $\gamma K^1 \cap K^1 \neq \varnothing$. Thus γ is in CC^{-1} which is compact. The group Γ being discrete, γ is in a finite set. $\square$

Exercise 2.10. Prove that if Γ is a Fuchsian group, then the orbits of Γ on $\mathbb{H}$ are closed and discrete, and the topological quotient space $\Gamma\backslash\mathbb{H}$ is separable. (Hint: [33, Theorem I.6.7].)

A Fuchsian group Γ *tessellates* $\mathbb{H}$, in the sense that there is a subset F of $\mathbb{H}$ satisfying the following conditions.

(i) F is a closed connected subset of $\mathbb{H}$ with non-empty interior;
(ii) $\bigcup_{\gamma \in \Gamma} \gamma F = \mathbb{H}$;
(iii) $\overset{\circ}{F} \cap \gamma \overset{\circ}{F} = \varnothing$, for all $\gamma \in \Gamma - \{\text{Id}\}$.

Such a set is called a *fundamental domain*. One established method for obtaining such domains is to choose a point z_0 of $\mathbb{H}$ which is not fixed by each element of $\Gamma - \{\text{Id}\}$ and to associate to z_0 the intersection of the half-planes (Figs. I.20 and I.21)

$$\mathbb{H}_{z_0}(\gamma) = \{z \in \mathbb{H} \mid d(z, z_0) \leqslant d(z, \gamma(z_0))\}.$$

The half-plane $\mathbb{H}_{z_0}(\gamma)$ is bounded by the perpendicular bisector $M_{z_0}(\gamma)$ of the hyperbolic segment $[z_0, \gamma(z_0)]_h$. Notice that, using the notation introduced just before Property 2.7, we have $\mathbb{H}_{z_0}(\gamma) = \mathbb{H} - \overset{\circ}{D}_{z_0}(\gamma)$.

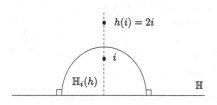

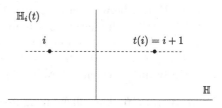

Fig. I.20. $h(z) = 2z$ **Fig. I.21.** $t(z) = z + 1$

Since the perpendicular bisector of the segment $[z_0, \gamma(z_0)]_h$ is a geodesic, the set $\mathbb{H}_{z_0}(\gamma)$ is *convex* (i.e., for any z and z' in $\mathbb{H}_{z_0}(\gamma)$, the hyperbolic segment $[z, z']_h$ is included in this set). It follows that the intersection of all $\mathbb{H}_{z_0}(\gamma)$ is also convex.

Define

$$\mathcal{D}_{z_0}(\Gamma) = \bigcap_{\substack{\gamma \in \Gamma \\ \gamma \neq \text{Id}}} \mathbb{H}_{z_0}(\gamma).$$

This set is called the *Dirichlet domain* of Γ centered at z_0.

Theorem 2.11. *A Dirichlet domain is a convex fundamental domain of Γ.*

Exercise 2.12. Prove Theorem 2.11.
(Hint: [41, Theorem 3.2.2].)

Let us examine the boundary of the Dirichlet domain $\mathcal{D}_{z_0}(\Gamma)$ in $\mathbb{H}$.

Property 2.13. *The set $\mathcal{D}_{z_0}(\Gamma) - \overset{\circ}{\mathcal{D}}_{z_0}(\Gamma)$ is in the union of the perpendicular bisectors of the segments $[z_0, \gamma(z_0)]_h$, with γ in $\Gamma - \{\text{Id}\}$.*

Proof. Take z' in $\mathcal{D}_{z_0}(\Gamma) - \overset{\circ}{\mathcal{D}}_{z_0}(\Gamma)$. By definition, z' is a limit of a sequence of points in $\mathcal{D}_{z_0}(\Gamma)$, and of points $(z_n)_{n \geqslant 1}$ in $\mathbb{H}$ satisfying $d(z_0, z_n) > d(z_0, \gamma_n(z_n))$ for some γ_n in Γ. It follows that the sequence $(\gamma_n^{-1}(z_0))_{n \geqslant 1}$ is bounded. Since Γ is discrete, the set of γ_n is finite and hence for some $\gamma' \neq \text{Id}$ in Γ, we have $d(z_0, z') \geqslant d(z_0, \gamma'(z'))$. Since z' is in $D_z(\Gamma)$, the point z' is in the perpendicular bisector of $[z_0, \gamma'(z_0)]_h$. $\square$

Corollary 2.14. *The hyperbolic area of $\mathcal{D}_{z_0}(\Gamma) - \overset{\circ}{\mathcal{D}}_{z_0}(\Gamma)$ is zero.*

Some fundamental domains of Γ may be topologically very wild (see [7, Example 9.2.5]). This is not the case with Dirichlet domains.

Property 2.15. *A Dirichlet domain $\mathcal{D}_z(\Gamma)$ is locally finite (i.e., for any compact K in $\mathbb{H}$, the set of γ in Γ satisfying $\gamma \mathcal{D}_z(\Gamma) \cap K \neq \varnothing$ is finite).*

Proof. Suppose not. Then for some compact subset K of $\mathbb{H}$ and some infinite subset Γ' of Γ, given any γ in Γ' the statement $\gamma \mathcal{D}_z(\Gamma) \cap K \neq \varnothing$ holds. The set Γ' can be written as a sequence $(\gamma_n)_{n \geqslant 1}$ in Γ. For any $n \geqslant 1$, there is $z_n \in \mathcal{D}_z(\Gamma)$ such that $\gamma_n(z_n)$ is in K. Since z_n is in $\mathcal{D}_z(\Gamma)$, one has that $d(z_n, z) \leqslant d(\gamma_n(z_n), z)$. From this inequality, one obtains that the sequence $(\gamma_n(z))_{n \geqslant 1}$ is bounded. The group Γ being Fuchsian, the set $\{\gamma_n \mid n \geqslant 1\}$ is necessarily finite, which contradicts our initial assumption. □

Examples 2.16. The following figures are examples of Dirichlet domains centered at i and associated to Fuchsian groups generated by a positive isometry g (Figs. I.22 ($g(z) = 2z$), I.23 ($g(z) = z + 1$) and I.24 (g is elliptic and does not fix i)).

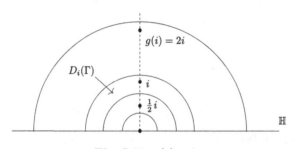

Fig. I.22. $g(z) = 2z$

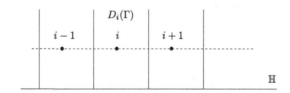

Fig. I.23. $g(z) = z + 1$

The following result, proved in [7, Theorem 9.2.4], shows that Dirichlet domains allow us to visualize the surface $S = \Gamma \backslash \mathbb{H}$ associated with Γ. More

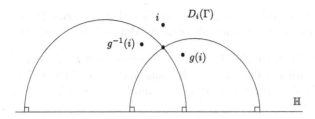

Fig. I.24. g elliptic

precisely, let $\Gamma\backslash\mathcal{D}_{z_0}(\Gamma)$ be the set of elements of $\mathcal{D}_{z_0}(\Gamma)$ modulo Γ, the function $\theta : \Gamma\backslash\mathcal{D}_{z_0}(\Gamma) \to S$ defined by

$$\theta(\Gamma z' \cap \mathcal{D}_{z_0}(\Gamma)) = \Gamma z',$$

satisfies

Proposition 2.17 ([7, Theorem 9.2.4]). *The map* $\theta : \Gamma\backslash\mathcal{D}_{z_0}(\Gamma) \to S$ *is a homeomorphism.*

As applications of this previous proposition, we can draw the surfaces associated to discrete cyclic groups.

Examples 2.18. The Figs. I.25 show surfaces associated to discrete cyclic groups introduced in Example 2.16. The first two surfaces are topologically

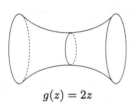

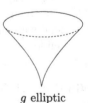

$g(z) = 2z$ $g(z) = z + 1$ g elliptic

Fig. I.25.

equivalent. Both are homeomorphic to a cylinder. From the metric point of view, however, there are some differences. To highlight these differences, choose a generator g of Γ. Define the curve $c = [g^{-1}(i), i]_h \cup [i, g(i)]_h$. Let c' be the intersection of c with the Dirichlet domain $\mathcal{D}_i(\Gamma)$. Now (referring to Fig. I.25) remove c' from $\mathcal{D}_i(\Gamma)$. In the first case, the curve c is the segment $[2i, 1/2i]_h$, thus c' splits the domain into two subdomains of infinite area.

In the second case, the isometry g is a translation, the curve c' which is not a geodesic segment splits the domain into two subdomains, one with finite area and the other of infinite area.

Observe that Proposition 2.17 implies that if some Dirichlet domain $\mathcal{D}_{z_0}(\Gamma)$ is compact in $\mathbb{H}$, then all Dirichlet domains of Γ are also compact.

Exercise 2.19. Prove that if the area of one domain $\mathcal{D}_z(\Gamma)$ is finite, then the area of any Dirichlet domain of Γ is finite and is equal to $\mathcal{A}(\mathcal{D}_z(\Gamma))$.
(Hint: use the fact that the area of the boundary of a Dirichlet domain is zero (Corollary 2.14).)

Definition 2.20. A Fuchsian group Γ is called a *lattice* if the area of each (or of one) Dirichlet domain $\mathcal{D}_z(\Gamma)$ is finite. Moreover, a lattice is said to be uniform if each (or one) Dirichlet domain is compact.

3 Limit points of Fuchsian groups

In this section, we will focus on the action of a Fuchsian group Γ on $\mathbb{H}(\infty)$. When Γ is not elementary, we associate to it a constellation of points in $\mathbb{H}(\infty)$, which will play a crucial role in the next chapters.

3.1 Limit set

Let us first analyze the Γ-orbit of a point z in $\mathbb{H}$. If Γ is infinite, since its action on $\mathbb{H}$ is properly discontinuous, Γz accumulates on $\mathbb{H}(\infty)$. In particular, there is a sequence $(\gamma_n(z))_{n \geqslant 1}$ of Γz converging to a point x in $\mathbb{H}(\infty)$. This point does not depend on z. To prove it, we can assume that $x = 0$. Set $\gamma_n(z) = a_n + ib_n$. For all z' in $\mathbb{H}$, we have $d(\gamma_n(z'), \gamma_n(z)) = d(z, z')$, hence $\gamma_n(z')$ is in the hyperbolic circle centered at $\gamma_n(z)$ with ray $d(z, z')$. This circle is the Euclidean circle having $[a_n + ib_n e^{d(z,z')}, a_n + ib_n e^{-d(z,z')}]$ as its diameter. The Euclidean length of this diameter tends to 0 and $\gamma_n(z)$ tends to 0, thus $(\gamma_n(z'))_{n \geqslant 1}$ also converges to 0.

Definition 3.1. The *limit set* $L(\Gamma)$ of Γ is the closed—possibly empty—subset of $\mathbb{H}(\infty)$ defined by

$$L(\Gamma) = \overline{\Gamma z} \cap \mathbb{H}(\infty).$$

This set does not depend on our choice of z and is Γ-invariant.

Exercise 3.2. Prove that if γ is a non-elliptic isometry of Γ, then $L(\Gamma)$ contains the fixed point(s) of γ.

The following proposition relates the existence of hyperbolic isometries in Γ to the cardinality of $L(\Gamma)$.

Proposition 3.3. *If Γ contains at least two hyperbolic isometries that do not have a fixed point in common, then $L(\Gamma)$ contains infinitely many elements. Otherwise,*

- *if all hyperbolic isometries of Γ have the same axis, then $L(\Gamma)$ is reduced to the endpoints of that axis,*
- *if Γ contains no hyperbolic isometries, then either $L(\Gamma)$ is the empty set, or $L(\Gamma)$ is reduced to a single point and Γ is generated by a parabolic isometry fixing this point.*

Proof. Suppose that Γ contains two hyperbolic isometries h_1, h_2 that do not have a fixed point in common. For $n \geqslant 1$, consider the transformation $g_n = h_1^n h_2 h_1^{-n}$. Each g_n belongs to G and is hyperbolic. Moreover the fixed points of g_n are the image by h_1^n of the fixed points of h_2. Since h_1, h_2 do not have a fixed point in common, all g_n are distinct. These fixed points are in $L(\Gamma)$ (Exercise 3.2).

Suppose now that all hyperbolic isometries of Γ have the same axis. Let x and y denote the endpoints of this axis. These two points are in $L(\Gamma)$ and are fixed by a hyperbolic isometry h of Γ. For all γ in Γ, the isometry $\gamma h \gamma^{-1}$ fixes x and y thus γ preserves the geodesic (xy), which shows that these two points are the only elements of $L(\Gamma)$.

Finally we consider the remaining case where Γ does not contain any hyperbolic isometries. Suppose that it contains a parabolic isometry p. After conjugating Γ, we can restrict our attention to the case where $p(z) = z + 1$. Let γ in Γ, we have $\gamma(\infty) = \infty$. Actually, if $\gamma(\infty) \neq \infty$, then for n large enough $|\mathrm{tr}(p^n \gamma)| > 2$ and hence $p^n \gamma$ is hyperbolic which is impossible. It follows that the group Γ is in the group of parabolic transformations of the form $z + b$ where b is a real number. Since Γ is discrete, it is generated by a translation and hence $L(\Gamma)$ is reduced to the point ∞.

It remains to analyze the case in which Γ only contains elliptic isometries. If all its elements are of order 2, then Γ is abelian and cyclic of order 2. Thus Γ fixes a single point of $\mathbb{H}$. If Γ is not cyclic of order 2, use the Poincaré disk model instead. After conjugating Γ, one may assume that it contains an isometry of the form $r(z) = e^{i\theta} z$ with $\theta \neq k\pi$. Let $g \in \Gamma$ written as $g(z) = (az + b)/(\bar{b}z + \bar{a})$ with $|a|^2 - |b|^2 = 1$. Calculating the trace of $rgr^{-1}g^{-1}$, one finds

$$\mathrm{tr}(rgr^{-1}g^{-1}) = 2 + 4|b|^2 \sin^2 \theta.$$

Since $rgr^{-1}g^{-1}$ is elliptic (or trivial), Property 2.5 implies that $b = 0$ and thus $g(0) = 0$. Consequently Γ is a subgroup of elliptic isometries fixing 0 and hence $L(\Gamma) = \varnothing$. $\qquad\square$

Exercise 3.4. Prove that if $L(\Gamma)$ is reduced to two points, then either Γ is generated by a hyperbolic isometry, or Γ contains a subgroup of index 2 of this form.

We deduce from Proposition 3.3 and Exercise 3.4 that, if the limit set of a Fuchsian Γ group is finite, then this set contains at most 2 points.

Definition 3.5. A Fuchsian group is said to be *elementary* if its limit set is finite.

Proposition 3.6. *If Γ is not elementary, then $L(\Gamma)$ is minimal, in the sense that $L(\Gamma)$ is the smallest (ordered by inclusion) non-empty, closed subset of $\mathbb{H}(\infty)$ which is Γ-invariant.*

Proof. Let F be a closed, non-empty, Γ-invariant subset of $L(\Gamma)$. Since Γ is not elementary, it contains infinitely many hyperbolic isometries $(\gamma_n)_{n \geqslant 1}$ having no shared fixed points (Proposition 3.3). Since the closed set F is invariant with respect to each γ_n, the fixed points γ_n^+ and γ_n^- are necessarily contained in F. Fix a positive integer N and choose a point z on the geodesic $(\gamma_N^- \gamma_N^+)$. Let x be a point in $L(\Gamma)$ and let $(g_n)_{n \geqslant 1}$ be a sequence in Γ such that $\lim_{n \to +\infty} g_n(z) = x$. Passing to a subsequence, one may assume that the sequences $(g_n(\gamma_N^-))_{n \geqslant 1}$ and $(g_n(\gamma_N^+))_{n \geqslant 1}$ converge to f^- and f^+ in F. Since $g_n(z)$ is in the geodesic $(g_n(\gamma_N^-)g_n(\gamma_N^+))$, the point x is necessarily in $\{f^-, f^+\}$. This shows that $L(\Gamma)$ is contained in F and thus $F = L(\Gamma)$. $\qquad \square$

Exercise 3.7. Prove that if Γ is not elementary, then none of the points in $L(\Gamma)$ is isolated.

Note that if Γ is not elementary, then $L(\Gamma)$ is uncountable since this set is closed, non-empty and contains no isolated points (Baire Category Theorem).

Exercise 3.8. Prove that if $L(\Gamma)$ differs from $\mathbb{H}(\infty)$, then it is totally discontinuous.
(Hint: use the density in $L(\Gamma)$ of the orbit of any point in $L(\Gamma)$.)

3.2 Horocyclic, conical and parabolic points

Different types of points in $L(\Gamma)$ may be distinguished by the ways in which they are approached by sequences in Γz.

Let us start with a point in $L(\Gamma)$ which is an attractor γ^+ of a hyperbolic isometry γ in Γ. After conjugating Γ, one may assume that $\gamma^+ = \infty$ and that $\gamma(z) = \lambda z$, with $\lambda > 1$. Since $\mathrm{Im}(\gamma^n(z)) = \lambda^n \mathrm{Im}\, z$, for any $a > 0$ and for n large enough (Fig. I.26), we have $\mathrm{Im}(\gamma^n(z)) > \ln a$. It follows that the points $\gamma^n(z)$, for n large enough, belong to the horodisk centered at ∞ defined by $H_a^+ = \{z \mid \mathrm{Im}\, z \geqslant a\}$. Notice that this property does not depend on z. In conclusion if $x \in L(\Gamma)$ is fixed by a hyperbolic isometry, then for any $z \in \mathbb{H}$, the orbit Γz meets every horodisk centered at x.

More generally, we have the following definition.

Definition 3.9. A point x in $L(\Gamma)$ is *horocyclic* if for any (or for one) z in $\mathbb{H}$, its orbit Γz meets every horodisk centered at x.

The set of such points is denoted by $L_h(\Gamma)$. In the course of our discussion we will show that, if Γ is not elementary, then $L_h(\Gamma)$ is uncountable (Lemma II.1.2 and Corollary II.1.7). Since Γ is countable, this property implies that most of the points in $L_h(\Gamma)$ are not fixed points of hyperbolic isometries of Γ.

Horocyclic points can be characterized in terms of Busemann cocycles.

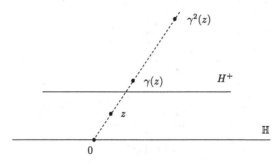

Fig. I.26.

Proposition 3.10. *A point x in $L(\Gamma)$ is horocyclic if and only if for any z in $\mathbb{H}$, we have $\sup_{\gamma \in \Gamma} B_x(z, \gamma(z)) = +\infty$.*

Proof. Let $x \in L(\Gamma)$, after conjugating Γ, one may assume that $x = \infty$. By definition, ∞ is horocyclic if for any horodisk $H_a^+ = \{z \mid \operatorname{Im} z \geqslant a\}$, with $a > 0$, there exists γ in Γ such that $\gamma(i)$ is in $H_a^+(x)$. This assertion is equivalent to the fact that for any $n \geqslant 1$, there exists γ_n in Γ such that $\operatorname{Im} \gamma_n(i) \geqslant n$. Since $B_\infty(i, z) = \ln(\operatorname{Im} z)$, we obtain that the point ∞ is horocyclic if and only if there exists a sequence $(\gamma_n)_{n \geqslant 1}$ in Γ such that $\lim_{n \to +\infty} B_\infty(i, \gamma_n(i)) = +\infty$. We achieve the proof using the fact that $B_\infty(z, \gamma_n(z)) = B_\infty(z, i) + B_\infty(i, \gamma_n(i)) + B_\infty(\gamma_n(i), \gamma_n(z))$ and $B_\infty(\gamma_n(i), \gamma_n(z)) \leqslant d(i, z)$. ▫

Some horocyclic points x are distinguished by having a sequence in Γz which approaches x along a path which remains within a bounded distance of the geodesic ray $[z, x)$.

Definition 3.11. A point x in $L(\Gamma)$ is *conical* if for some z in $\mathbb{H}$, there exist $\varepsilon > 0$ and $(\gamma_n)_{n \geqslant 0}$ in Γ such that the sequence $(\gamma_n(z))_{n \geqslant 0}$ converges to x and $d(\gamma_n(z), [z, x)) \leqslant \varepsilon$.

Notice that the fixed point of a hyperbolic isometry γ of Γ is conical since, if z belongs to the axis of γ, then the sequence $(\gamma^n(z))_{n \geqslant 1}$ is in this axis.

The following exercise shows that, if x is conical, then for any z in $\mathbb{H}$, there is an infinite subsequence at a bounded distance of the ray $[z, x)$.

Exercise 3.12. Let x in $L(\Gamma)$ and $(\gamma_n)_{n \geqslant 0}$ in Γ. Prove that, if for some $z \in \mathbb{H}$ the sequence $(\gamma_n(z))_{n \geqslant 0}$ remains at a bounded distance of $[z, x)$, then the same property holds for any z' in $\mathbb{H}$.

The set of conical points is denoted $L_c(\Gamma)$. Clearly we have:

$$L_c(\Gamma) \subset L_h(\Gamma).$$

The terminology "conical" arises from the shape of ε-neighborhoods of vertical geodesics.

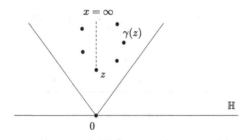

Fig. I.27.

Exercise 3.13. Let $\varepsilon > 0$. Prove that there is some $\varepsilon' > 0$ such that: $d(z, (0\infty)) \leqslant \varepsilon$ if and only if $|\operatorname{Re} z / \operatorname{Im} z| \leqslant \varepsilon'$ (Fig. I.27). (Hint: use Exercise 1.8.)

Conical points can be characterized in terms of the action of Γ on the product $\mathbb{H}(\infty) \times \mathbb{H}(\infty)$.

Proposition 3.14. *A point x in $L(\Gamma)$ is conical if and only if there exists a sequence $(\gamma_n)_{n\geqslant 0}$ of different transformations of Γ such that, for all y in $\mathbb{H}(\infty)$ different from x, the sequence $(\gamma_n(x), \gamma_n(y))_{n\geqslant 0}$ remains in a compact subset of $\mathbb{H}(\infty) \times \mathbb{H}(\infty)$ with its diagonal removed.*

Proof. Take the Poincaré disk model and $x \neq y$ in $\mathbb{D}(\infty)$. Clearly a sequence $(\gamma_n(x), \gamma_n(y))_{n\geqslant 0}$ remains in a compact subset of $\mathbb{D}(\infty) \times \mathbb{D}(\infty)$ with its diagonal removed, if and only if the set of the hyperbolic distances between the origin 0 of $\mathbb{D}$ and the geodesics $(\gamma_n(x)\gamma_n(y))_{n\geqslant 0}$ is bounded. This condition is equivalent to the fact that the set of points $\gamma_n^{-1}(0)$ is included in some ε-neighborhood of the geodesic (xy). Notice that the point 0 can be replaced by any $z \in \mathbb{D}$.

Suppose that x is a conical point in $L(\Gamma)$. For any z in $\mathbb{D}$, there exist $\varepsilon > 0$ and $(\gamma_n)_{n\geqslant 0}$ in Γ such that the sequence $(\gamma_n(z))_{n\geqslant 0}$ converges to x and $d(\gamma_n(z), [z, x)) \leqslant \varepsilon$. Take y in $\mathbb{D}(\infty)$ different from x, and z' in the geodesic (xy). The sequence $d(\gamma_n(z'), [z', x))$ is also bounded, and hence the point z' remains at a bounded distance from the sequence of geodesics $(\gamma_n^{-1}(y)\gamma_n^{-1}(x))$. It follows that the sequence $(\gamma_n^{-1}(x), \gamma_n^{-1}(y))_{n\geqslant 0}$ remains in a compact subset of $\mathbb{D}(\infty) \times \mathbb{D}(\infty)$ with its diagonal removed.

Suppose now that there exists a sequence $(\gamma_n)_{n\geqslant 0}$ of different transformations of Γ such that, for all y in $\mathbb{H}(\infty)$ different from x, the sequence $(\gamma_n(x), \gamma_n(y))_{n\geqslant 0}$ remains in a compact subset of $\mathbb{H}(\infty) \times \mathbb{H}(\infty)$ with its diagonal removed. It follows that the points $\gamma_n^{-1}(0)$ remain at a bounded distance from the geodesic (xy). Since all γ_n are different, the set S of points $\gamma_n^{-1}(0)$ accumulates to $\mathbb{D}(\infty)$. Translating Exercise 3.13 in $\mathbb{D}$, one obtains that the closure of S in $\mathbb{D} \cup \mathbb{D}(\infty)$ is included in $\{x, y\}$, for any $y \neq x$, and hence that the sequence $(\gamma_n^{-1}(0))_{n\geqslant 0}$ converges to x. $\qquad\square$

Later in the text, we will assume a regularity hypothesis on the group Γ (see Sect. 4), namely equality of $L_h(\Gamma)$ and $L_c(\Gamma)$. In general, this equality does not hold (in an article by A. Starkov [59], the reader will find some examples of groups having horocyclic points which are not conical).

Conical points can be characterized in terms of distance and Busemann cocycles.

Proposition 3.15. *Let z be in $\mathbb{H}$. A point x in $L(\Gamma)$ is conical if and only if there is a sequence $(\gamma_n)_{n \geqslant 1}$ in Γ satisfying the following two conditions:*

(i) $\lim_{n \to +\infty} B_x(z, \gamma_n(z)) = +\infty$;
(ii) $(d(z, \gamma_n(z)) - B_x(z, \gamma_n(z)))_{n \geqslant 1}$ *is a bounded sequence.*

Proof. After conjugating Γ, one can recover the case in which $z = i$ and $x = \infty$.

Suppose that the point ∞ is conical. Then there are some $\varepsilon > 0$ and some sequences $(\gamma_n)_{n \geqslant 1}$ in Γ and $(t_n)_{n \geqslant 1}$ in $\mathbb{R}^+$ such that

$$\lim_{n \to +\infty} t_n = +\infty \quad \text{and} \quad d(it_n, \gamma_n(i)) \leqslant \varepsilon.$$

In particular, we have $\lim_{n \to +\infty} B_x(z, \gamma_n(z)) = +\infty$. Moreover, the point $\gamma_n(i)$ is in the Euclidean disk having $[it_n e^\varepsilon, it_n e^{-\varepsilon}]$ as a diameter, therefore the point $z'_n = i \operatorname{Im} \gamma_n(i)$ is also in this disk. For large enough integers n, we have $B_\infty(i, \gamma_n(i)) = d(i, z'_n)$; therefore $|d(i, \gamma_n(i)) - B_\infty(i, \gamma_n(i))| \leqslant d(\gamma_n(i), z'_n)$. This inequality implies that $|d(i, \gamma_n(i)) - B_\infty(i, \gamma_n(i))|$ is less than 2ε.

Suppose now that there is a sequence $(\gamma_n)_{n \geqslant 1}$ in Γ satisfying the conditions of Proposition 3.15. Then for n large enough and for $z'_n = i \operatorname{Im} \gamma_n(i)$, we have

$$B_\infty(i, \gamma_n(i)) = d(i, z'_n),$$

and $\lim_{n \to +\infty} z'_n = \infty$.

Additionally, there is an $A > 0$ such that

$$|d(i, \gamma_n(i)) - \ln(\operatorname{Im} \gamma_n(i))| \leqslant A.$$

The sequence $(e^{1/2 d(i, \gamma_n(i))} / \sqrt{\operatorname{Im} \gamma_n(i)})_{n \geqslant 1}$ is therefore bounded above.

According to Exercise 1.8, the sequence $(|i - \gamma_n(i)| / \operatorname{Im} \gamma_n(i))_{n \geqslant 1}$ is likewise bounded above. Thus there exists $A' > 0$ such that

$$\left| \frac{\operatorname{Re} \gamma_n(i)}{\operatorname{Im} \gamma_n(i)} \right| \leqslant A'.$$

We deduce from Exercise 3.13, that the sequence $(\gamma_n(i))_{n \geqslant 1}$ is in an ε-neighborhood of $[i, \infty)$. $\qquad\square$

The last type of limit points that we introduce in this text are *parabolic* points.

Definition 3.16. A point x in $L(\Gamma)$ is *parabolic* if there exists a parabolic transformation $\gamma \neq \mathrm{Id}$ in Γ such that $\gamma(x) = x$. The set of parabolic points is denoted $L_p(\Gamma)$.

Recall that, if x is fixed by a parabolic isometry γ in Γ, then the sequence $(\gamma^n(z))_{n \geqslant 1}$ converges to x along the horocycle centered at x passing through z.

Is it possible for a parabolic point to be horocyclic? To answer this question, we will prove a stronger result.

Theorem 3.17. *Let x be a point in $L_p(\Gamma)$. There exists a horodisk $H^+(x)$ centered at x such that*

$$\gamma H^+(x) \cap H^+(x) = \varnothing,$$

for any γ in Γ which does not fix x (Fig. I.28).

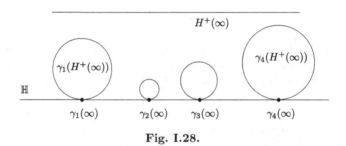

Fig. I.28.

Before we prove this theorem, notice that if x is parabolic and z is chosen in a horodisk $H^+(x)$ given by Theorem 3.17, then Γz does not meet any horodisks centered at x which is properly contained in $H^+(x)$. Therefore x cannot be horocyclic.

Corollary 3.18. *A parabolic point is not horocyclic.*

It is sufficient to prove Theorem 3.17 in the case where $x = \infty$. In the proof, we will use the following two results on the Euclidean diameter of the image of a horodisk centered at ∞ under some isometry which does not fix the point ∞.

Lemma 3.19. *Let $g(z) = (az + b)/(cz + d)$ be a Möbius transformation, where a, b, c, d are real numbers satisfying $c \neq 0$ and $ad - bc = 1$. For $t > 0$, consider the horodisk H^+ centered at ∞ defined by $H^+ = \{z \in \mathbb{H} \mid \mathrm{Im}\, z \geqslant t\}$. The horodisk $g(H^+)$ is the Euclidean disk tangent to the real axis at the point a/c, with Euclidean diameter $1/c^2 t$.*

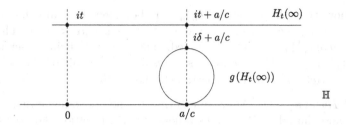

Fig. I.29.

Proof. Since $c \neq 0$, the isometry g sends H^+ onto the Euclidean disk tangent to the real axis at the point $g(\infty) = a/c$. Moreover the point $g(it)$ is in the horocycle $g(H)$ associated to $g(H^+)$ (Fig. I.29 $H_t(\infty) = H$). Let δ denote the Euclidean diameter of $g(H)$. By definition of the Busemann cocycle, we have

$$B_{a/c}\,(it + a/c, i\delta + a/c) = \ln t/\delta$$

and thus, $B_\infty(g^{-1}(it + a/c), g^{-1}(i\delta + a/c)) = \ln t/\delta$.

Additionally, the point $i\delta + a/c$ is in $g(H)$, hence $g^{-1}(i\delta + a/c)$ is in H. Since it is also in H, we have $B_\infty(it, g^{-1}(i\delta + a/c)) = 0$ and hence

$$B_\infty(g^{-1}(it + a/c), g^{-1}(i\delta + a/c)) = B_\infty(g^{-1}(it + a/c), it).$$

Therefore,

$$\ln t/\delta = \ln \frac{t}{\operatorname{Im} g^{-1}(it + a/c)},$$

which implies the equality $\delta = 1/c^2(g)t$.   □

Notation. Let $g(z) = (az + b)/(cz + d)$ be a Möbius transformation, where a, b, c, d are real numbers satisfying $ad - bc = 1$. The positive real number $|c|$ is denoted $c(g)$.

Notice that $c(g) = 0$ if and only if g fixes the point ∞. Moreover if g is a translation, then $c(g^n h g^m) = c(h)$ for all n, m and $h \in G$.

Proposition 3.20. *If Γ contains a non-trivial translation, then there exists $A > 0$, such that $c(\gamma) \geqslant A$ for all γ in Γ satisfying $\gamma(\infty) \neq \infty$.*

Proof. Let $g(z) = z + \alpha$, where $\alpha > 0$, be a translation in Γ. Suppose that there is a sequence $(\gamma_n)_{n \geqslant 1}$ in Γ such that $\gamma_n(\infty) \neq \infty$ and $\lim_{n \to +\infty} c(\gamma_n) = 0$. Write γ_n in the form $\gamma_n(z) = (a_n z + b_n)/(c_n z + d_n)$ with $a_n d_n - b_n c_n = 1$ and $c_n > 0$.

Consider the integer parts, e_n and e'_n, of $a_n/\alpha c_n$ and $d_n/\alpha c_n$ respectively. Setting $g_n = g^{-e'_n} \gamma_n g^{-e_n}$, one has

$$g_n(z) = \frac{(a_n - c_n e_n \alpha)z + b'_n}{c_n z + (-c_n e'_n \alpha + d_n)}.$$

We know that the sequence $(c_n)_{n\geqslant 1}$ is bounded. One also has the inequalities $0 < a_n - c_n e_n \alpha < c_n \alpha$ and $0 < d_n - c_n e'_n \alpha < c_n \alpha$. Thus the sequence $(g_n g g_n^{-1})_{n\geqslant 1}$ is also bounded. The group Γ being discrete, the set $\{g_n g g_n^{-1} \mid n \geqslant 1\}$ is finite. As a result, in the tail of the sequence, $c(g_n g g_n^{-1}) = \alpha c(\gamma_n)^2 = 0$. This contradicts the hypothesis $\gamma_n(\infty) \neq \infty$. $\qquad\square$

Proof (of Theorem 3.17). After conjugating Γ, one may assume that x is the point ∞, and thus that Γ contains a non-trivial translation. Following from Proposition 3.20, there exists $A > 0$ such that $c(\gamma) > A$ for all γ in Γ which do not fix the point ∞. Take $t = 2/A$. From Lemma 3.19, such γ sends the horodisk $H^+ = \{z \in \mathbb{H} \mid \operatorname{Im} z \geqslant t\}$ onto an Euclidean disk having Euclidean diameter $A/2c^2(\gamma)$, tangent to the real axis. Since $c(\gamma) > A$, this diameter is less than $1/2A$, hence gH^+ does not meet H^+. $\qquad\square$

Let y be a point in $\overline{\mathbb{H}}$ and denote by Γ_y the subgroup of Γ fixing y.

Exercise 3.21. Prove that Γ_y is cyclic.

If x is a point in $L_p(\Gamma)$, then the group Γ_x is generated by a parabolic isometry and hence preserves each horodisk centered at x. Fix a horodisk $H^+(x)$ given by Theorem 3.17.

Lemma 3.22. *The natural projection $q : \Gamma_x \backslash H^+(x) \to \Gamma \backslash \mathbb{H}$ is injective.*

Proof. Take y and y' in $H^+(x)$ and suppose that $q(\Gamma_x(y)) = q(\Gamma_x(y'))$. There exists γ in Γ such that $y' = \gamma(y)$. Hence $H^+(x) \cap \gamma H^+(x) \neq \varnothing$. It follows from Theorem 3.17 that γ is in Γ_x and hence, that $\Gamma_x(y) = \Gamma_x(y')$. $\qquad\square$

Definition 3.23. The subset $q(\Gamma_x \backslash H^+(x))$ of $\Gamma \backslash \mathbb{H}$ is called the *cusp* associated with $H^+(x)$ and is denoted by $C(H^+(x))$.

To visualize the shape of a cusp, without loss of generality we can suppose $x = \infty$, $H^+(\infty) = \{z \in \mathbb{H} \mid \operatorname{Im} z \geqslant t\}$ with $t > 0$, and Γ_x is generated by a translation $g(z) = z + a$, with $a > 0$. Fix a point $z \in \mathbb{H}$ such that $\operatorname{Re} z = a/2$ and that $\gamma(z) \neq z$ for any $\gamma \neq \operatorname{Id}$ in Γ. Applying Proposition 2.17 to the Dirichlet domains $\mathcal{D}_z(\Gamma_x)$ and $\mathcal{D}_z(\Gamma)$, we obtain that the cusp $C(H^+(\infty))$ is homeomorphic to the vertical strip $B = \{z' \in \mathbb{H} \mid 0 \leqslant \operatorname{Re}(z') \leqslant a, \operatorname{Im} z' \geqslant t\}$ (Fig. I.30) in which the vertical edges are identified by the translation g (Fig. I.31).

4 Geometric finiteness

Recall that a lattice is a Fuchsian group Γ such that the area of each (or of one) Dirichlet domain $\mathcal{D}_z(\Gamma)$ is finite, and that a lattice is said to be uniform if each (or one) Dirichlet domain is compact (Definition 2.20). The purpose of this section is to generalize these notions by adopting two points of view: one involving the geometry of a Dirichlet domain, the other involving the properties of points in the limit set.

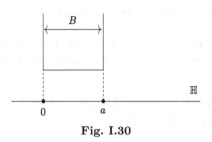

Fig. I.30

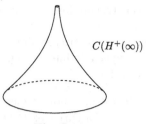

Fig. I.31

4.1 Geometric finiteness and Dirichlet domains

Let Γ be a Fuchsian group. We associate to it the following subset of $\mathbb{H}$ defined by:

$$\widetilde{\Omega}(\Gamma) = \{z \in \mathbb{H} \mid \text{there exist } x \text{ and } y \text{ in } L(\Gamma) \text{ such that } z \in (xy)\}.$$

If Γ is elementary, then from Proposition 3.3, either $\widetilde{\Omega}(\Gamma)$ is empty or this set is a single geodesic.

Definition 4.1. Let Γ be a non-elementary Fuchsian group, the *Nielsen region* of Γ is the convex hull of $\widetilde{\Omega}(\Gamma)$, in the sense of hyperbolic segments. It is denoted by $N(\Gamma)$.

Exercise 4.2. Let Γ be a non-elementary Fuchsian group, prove that its Nielsen region is the smallest closed, non-empty and Γ-invariant subset of $\mathbb{H}$.

Proposition 4.3. *We have $N(\Gamma) = \mathbb{H}$ if and only if $L(\Gamma) = \mathbb{H}(\infty)$.*

Proof. If $L(\Gamma) = \mathbb{H}(\infty)$, then $N(\Gamma) = \mathbb{H}$. Suppose now that $L(\Gamma) \neq \mathbb{H}(\infty)$, and let us prove that $N(\Gamma)$ is properly contained in $\mathbb{H}(\infty)$. Since $L(\Gamma) \neq \mathbb{H}(\infty)$, there exists an open, non-empty interval I in $\mathbb{R}$ which has no intersection with $L(\Gamma)$. Clearly, the set $\widetilde{\Omega}(\Gamma)$ is included in the closed half-plane bounded by the geodesic whose endpoints are those of I, and whose boundary at infinity is $\mathbb{H}(\infty) - I$. Since this half-plane is convex the Nielsen region of Γ is also included in it, and hence $N(\Gamma) \neq \mathbb{H}$. □

Notice that the interior of the Nielsen region of a non-elementary Fuchsian group is not empty.

Definition 4.4. A Fuchsian group Γ is said to be *geometrically finite* if, either Γ is elementary, or there is a Dirichlet $\mathcal{D}_z(\Gamma)$ such that the set $N(\Gamma) \cap \mathcal{D}_z(\Gamma)$ has finite area.

For example, lattices are geometrically finite.

Clearly, the Nielsen region $N(\Gamma)$ of a Fuchsian group Γ is tessellated by the images by Γ of $N(\Gamma) \cap \mathcal{D}_z(\Gamma)$, for any Dirichlet domain. Thus the action of a geometrically finite group Γ on $N(\Gamma)$ behaves like the action of a lattice on $\mathbb{H}$.

Among geometrically finite groups, those whose action on $N(\Gamma)$ is cocompact are especially interesting. The following definition generalizes the notion of uniform lattices.

Definition 4.5. A non-elementary Fuchsian group Γ is called *convex-cocompact* if there is a Dirichlet $\mathcal{D}_z(\Gamma)$ such that the set $N(\Gamma) \cap \mathcal{D}_z(\Gamma)$ is compact.

Can the geometric finiteness of a Fuchsian group Γ be checked directly from the shape of one of its Dirichlet domains?

Before we answer this question, let us introduce the notion of edges and vertices of a Dirichlet domain. Take a Dirichlet domain $\mathcal{D}_z(\Gamma)$ of Γ and a non-trivial transformation γ of Γ. When the intersection of $\mathcal{D}_z(\Gamma)$ and $\gamma(\mathcal{D}_z(\Gamma))$ is non-empty, it is contained in the perpendicular bisector $M_z(\gamma)$ of $[z, \gamma(z)]_h$. This intersection, is a point, a non-trivial geodesic segment, a geodesic ray or a geodesic. In the latter three cases, we say that this intersection is an *edge* and denote it by $C(\gamma)$:

$$C(\gamma) = \mathcal{D}_z(\Gamma) \cap \gamma \mathcal{D}_z(\Gamma).$$

Notice that $\gamma^{-1} C(\gamma)$ is also a edge since $\gamma^{-1} C(\gamma) = \gamma^{-1} \mathcal{D}_z(\Gamma) \cap \mathcal{D}_z(\Gamma)$. Moreover this set is included in the perpendicular bisector of $[z, \gamma^{-1}(z)]_h$, hence $\gamma^{-1} C(\gamma) = C(\gamma^{-1})$.

The *vertices* are the endpoints of the edges. An *infinite vertex* is a vertex contained in $\mathbb{H}(\infty)$.

The group Γ being countable, the set of edges, and hence the set of vertices, of $\mathcal{D}_z(\Gamma)$ is also countable. Let $(C_i = C(\gamma_i))_{i \in I}$ the (possibly finite) sequence of edges of $\mathcal{D}_z(\Gamma)$, where I is a subset of $\mathbb{N}$ and γ_i is in Γ.

Exercise 4.6. Prove that $\mathcal{D}_z(\Gamma)$ is the intersection of the closed half-planes associated to the edges $C(\gamma_i)_{i \in I}$ of this domain, defined by $\{z' \in \mathbb{H} \mid d(z', z) \leqslant d(z', \gamma_i(z))\}$, for all i in I.

Exercise 4.7. Suppose that I is finite and that Γ is not elementary. Prove that the area of $\mathcal{D}_z(\Gamma)$ is infinite if and only if the boundary at infinity of this domain, $\mathcal{D}_z(\Gamma)(\infty) = \overline{\mathcal{D}_z(\Gamma)} \cap \mathbb{H}(\infty)$, contains a closed interval whose endpoints are two distinct infinite vertices.

Suppose now that Γ is not elementary. Let us prove that, if the domain $\mathcal{D}_z(\Gamma)$ has finitely many edges, then Γ is geometrically finite.

If the area of this domain is finite, then Γ clearly is geometrically finite (it is a lattice).

If the area of the domain $\mathcal{D}_z(\Gamma)$ is infinite, applying Exercise 4.7, we obtain a closed interval J contained in $\mathcal{D}_z(\Gamma)(\infty)$ whose endpoints are two distinct infinite vertices of $\mathcal{D}_z(\Gamma)$. Consider the finite sequence $(J_j)_{1 \leqslant j \leqslant k}$ of all such intervals. These intervals are pairwise disjoint. By construction, the polygon $\mathcal{P}$ bounded by the geodesics L_j whose endpoints are those of J_j, and by the edges $(C_i = C(\gamma_i))_{i \in I}$ of $\mathcal{D}_z(\Gamma)$ has finite area. Since $\Gamma(z) \cap \mathcal{D}_z(\Gamma) = \{z\}$, the interior of each interval J_j does not meet $L(\Gamma)$. It follows that the set $\widetilde{\Omega}(\Gamma)$, and hence $N(\Gamma)$, is included in the intersections of the closed half-planes bounded by L_j, whose boundary at infinity is $\overline{\mathbb{H}(\infty) - J_j}$, with the closed half-planes bounded by C_i containing z. Consequently, $N(\Gamma) \cap \mathcal{D}_z(\Gamma)$ is a subset of the polygon $\mathcal{P}$ (Fig. I.32), and hence Γ is geometrically finite.

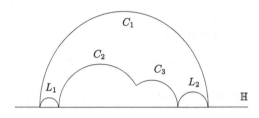

Fig. I.32.

The converse is also true. Its proof, which is more technical, requires a study of the vertices of a Dirichlet domain that we have decided not to include in the development of this text. Instead, we direct the reader to the proof by A. Beardon. These results are stated in the following theorem:

Theorem 4.8 ((i) $\Rightarrow$ (ii) [7, Theorem 10.1.2]). *Let Γ be a non-elementary Fuchsian group. Then the following are equivalent:*

(i) *The area of $\mathcal{D}_z(\Gamma) \cap N(\Gamma)$ is finite.*
(ii) *The Dirichlet domain $\mathcal{D}_z(\Gamma)$ has finitely many edges.*

4.2 Geometric finiteness and limit points

The goal of this subsection is to give a characterization of the geometric finiteness in terms of limit points.

Let us first analyze the intersection of the boundary at infinity of the Dirichlet domain $\mathcal{D}_z(\Gamma)$ of Γ with its limit set.

Proposition 4.9. *Let Γ be a Fuchsian group and $z \in \mathbb{H}$. A point x in $\mathbb{H}(\infty)$ is in $\mathcal{D}_z(\Gamma)(\infty)$ if and only if*

$$\sup_{\gamma \in \Gamma} B_x(z, \gamma(z)) = 0.$$

Proof. Let $z \in \mathbb{H}$ such that $\gamma(z) \neq z$ for any $\gamma \neq \mathrm{Id}$ in Γ and $x \in \mathbb{H}(\infty)$. Denote by $r : [0, +\infty) \to \mathbb{H}$ the arclength parametrization of the geodesic ray $[z, x)$.

Suppose that $\sup_{\gamma \in \Gamma} B_x(z, \gamma(z)) = 0$. To each γ in Γ we associate a function $f : \mathbb{R}^+ \to \mathbb{R}$ defined by $f(t) = t - d(\gamma(z), r(t))$. This is an increasing function since if $s > t$, one has $d(\gamma(z), r(s)) \leqslant d(\gamma(z), r(t)) + s - t$. Also, by definition of the Busemann cocycle, we have $\lim_{t \to +\infty} f(t) = B_x(z, \gamma(z))$. Since $B_x(z, \gamma(z))$ is negative, we obtain $f(t) = d(z, r(t)) - d(\gamma(z), r(t)) \leqslant 0$, for all $t \in \mathbb{R}^+$. This shows that the ray $[z, x)$ is included in $\mathcal{D}_z(\Gamma)$, and therefore that x is in $\mathcal{D}_z(\Gamma)(\infty)$.

Suppose now that x is in $\mathcal{D}_z(\Gamma)(\infty)$ and consider a sequence of points $(z_n)_{n \geqslant 0}$ in $\mathcal{D}_z(\Gamma)$ converging to x. Because $\mathcal{D}_z(\Gamma)$ is convex, one may assume that the sequence $(z_n)_{n \geqslant 0}$ is in the ray $[z, x)$. The point z_n is in $\mathcal{D}_z(\Gamma)$, thus for all γ in Γ one has $d(z, z_n) - d(\gamma(z), z_n) \leqslant 0$. Passing to the limit, one obtains $B_x(z, \gamma(z)) \leqslant 0$ for all γ in Γ, and thus $\sup_{\gamma \in \Gamma} B_x(z, \gamma(z)) = 0$. □

Geometrically, the previous proposition says that a point x in $\mathcal{D}_z(\Gamma)(\infty)$ is characterized by the fact that the orbit Γz does not intersect the interior of the horodisk centered at x passing through z.

Corollary 4.10. *Let Γ be a non-elementary Fuchsian group and x in $\mathbb{H}(\infty)$.*

- *If $x \notin L(\Gamma)$, then there exists $\gamma \in \Gamma$ such that $\gamma(x)$ is in $\mathcal{D}_z(\Gamma)(\infty)$.*
- *If $x \in L_p(\Gamma)$, then there exists $\gamma \in \Gamma$ such that $\gamma(x)$ is in $\mathcal{D}_z(\Gamma)(\infty)$. Moreover each point in $L_p(\Gamma) \cap \mathcal{D}_z(\Gamma)(\infty)$ is isolated in $\mathcal{D}_z(\Gamma)(\infty)$.*
- *The set $L_h(\Gamma) \cap \mathcal{D}_z(\Gamma)(\infty)$ is the empty set.*

Proof. Notice that the third property is a direct consequence of Propositions 4.9 and 3.10. Let us now prove the first two properties.

Let $g \in \Gamma$, applying Proposition 4.9 we obtain that for the point $g(x)$ is in $\mathcal{D}_z(\Gamma)(\infty)$ if and only if $\sup_{\gamma \in \Gamma} B_{g(x)}(z, \gamma(z)) = 0$. Since $B_{g(x)}(z, \gamma(z)) = B_x(g^{-1}z, z) + B_x(z, g^{-1}\gamma(z))$, we obtain that $g(x)$ is in $\mathcal{D}_z(\Gamma)(\infty)$ if and only if the number $S = \sup_{\gamma \in \Gamma} B_x(z, \gamma(z))$ is equal to $B_x(z, g^{-1}(z))$.

Notice that, if the point x is not in $L(\Gamma)$, or is in $L_p(\Gamma)$, then S is finite, because under these conditions x cannot be horocyclic (Corollary 3.18).

Suppose that there is some sequence $(\gamma_n)_{n \geqslant 0}$ in Γ such that the sequence $(B_x(z, \gamma_n(z)))_{n \geqslant 0}$ is not stationary and converges to S.

For n large enough, $\gamma_n(z)$ is in the horodisk $\{z' \in \mathbb{H} \mid B_x(z, z') \geqslant S - 1\}$. The intersection of this horodisk with $\mathbb{H}(\infty)$ is x. Therefore the sequence $(\gamma_n(z))_{n \geqslant 0}$ converges to x. If x does not belong to $L(\Gamma)$, we obtain a contradiction.

If x is parabolic, after conjugating Γ one may suppose that $x = \infty$. Since $(B_\infty(z, \gamma_n(z)))_{n \geqslant 0}$ is not stationary and converges to 0, one can choose A and B strictly positive such that for any n:

$$A \leqslant \operatorname{Im} \gamma_n(z) \leqslant B.$$

Let g be a non-trivial translation in Γ. For each n, there exists k_n for which the sequence $(g^{k_n} \gamma_n(z))_{n \geqslant 0}$ is bounded.

Since the group Γ is Fuchsian, the set $\{g^{k_n}\gamma_n(z) \mid n \geqslant 0\}$ is finite, which is impossible since $B_\infty(z, g^{k_n}\gamma_n(z)) = B_\infty(z, \gamma_n(z))$ and the sequence $(B_\infty(z, \gamma_n(z)))_{n \geqslant 0}$ is not stationary.

To prove the last part of the second property, take a point in $L_p(\Gamma) \cap \mathcal{D}_z(\Gamma)(\infty)$ and suppose $x = \infty$. Since the point ∞ is a parabolic point, the group Γ contains a non-trivial translation g, and hence $\mathcal{D}_z(\Gamma)$ is in the vertical strip bounded by the perpendicular bisectors of the segments $[z, g(z)]_h$ and $[z, g^{-1}(z)]_h$. It follows that the set $D_z(\Gamma)(\infty)$ with the point ∞ removed, is in a bounded interval of $\mathbb{R}$, which shows that the point ∞ is isolated in $D_z(\Gamma)(\infty)$. $\square$

Applying Corollary 4.10 to the particular case where the group Γ is a lattice, we obtain

Proposition 4.11. *If Γ is a lattice, then $L(\Gamma) = \mathbb{H}(\infty)$. Moreover the set of parabolic points $L_p(\Gamma)$ is empty or is the union of finitely many Γ-orbits.*

Proof. If Γ is a uniform lattice, then Proposition 4.11 is an immediate consequence of Corollary 4.10, since in this case $\mathcal{D}_z(\Gamma)$ is a compact subset of $\mathbb{H}$.

If Γ is a nonuniform lattice, then $\mathcal{D}_z(\Gamma)(\infty)$ is not empty, let us show that this set is finite.

Suppose that this is not the case and consider infinitely many points $(x_n)_{n \geqslant 0}$ in $\mathcal{D}_z(\Gamma)(\infty)$. Each x_n is the limit of a sequence of points in $\mathcal{D}_z(\Gamma)$. However, this domain is convex, thus the ray $[z, x_n)$ is a subset of $\mathcal{D}_z(\Gamma)$. Let T_n be the hyperbolic triangle with vertices z, x_n, x_{n+1}. After some re-ordering of $(x_n)_{n \geqslant 0}$, one may assume that these triangles T_n are adjacent. Let α_n denote the measure of the geometrical angle of T_n at z. The area of T_n is

$$\mathcal{A}(T_n) = \pi - \alpha_n.$$

Additionally, the triangles T_n being adjacent, one has $\sum_{n=0}^{+\infty} \alpha_n \leqslant 2\pi$. For all $N \geqslant 0$, the union of the triangles $\bigcup_{n=0}^{N} T_n$ is in $\mathcal{D}_z(\Gamma)$ and $\mathcal{A}(\bigcup_{n=0}^{N} T_n) \geqslant (N+1)\pi - 2\pi$, which is impossible since the area of $\mathcal{D}_z(\Gamma)$ is finite. In conclusion, the set $\mathcal{D}_z(\Gamma)(\infty)$ is finite. Applying Corollary 4.10, we obtain that the sets $\mathbb{H}(\infty) - L(\Gamma)$ and $L_p(\Gamma)$ are the finite union of Γ-orbits (or are empty). Since $\mathbb{H}(\infty) - L(\Gamma)$ is an open set, we have $\mathbb{H}(\infty) - L(\Gamma) = \varnothing$, hence $L(\Gamma) = \mathbb{H}(\infty)$. $\square$

More generally we have:

Theorem 4.12. *If Γ is a non-elementary geometrically finite Fuchsian group, then the set $L(\Gamma) \cap \mathcal{D}_z(\Gamma)(\infty)$ is finite, and is equal to the set $L_p(\Gamma) \cap \mathcal{D}_z(\Gamma)(\infty)$. Moreover the set $L_p(\Gamma)$ is a finite union of Γ-orbits.*

Proof. For $g \in G$ and z in $\mathbb{H}$ such that $g(z) \neq z$, recall that the closed half-plane defined by $D_z(g) = \{z' \in \mathbb{H} \mid d(z', z) \geqslant d(z', g(z))\}$ satisfies: $g(\mathring{D}_z(g^{-1})) = \mathbb{H} - D_z(g)$. We choose z and denote $D_z(g) = D(g)$.

The group Γ being geometrically finite, some Dirichlet domain $\mathcal{D}_z(\Gamma)$ has finitely many edges. We have $\Gamma(z) \cap \mathcal{D}_z(\Gamma) = \{z\}$, hence x belongs to $L(\Gamma) \cap \mathcal{D}_z(\Gamma)(\infty)$ and x is an infinite vertex of $\mathcal{D}_z(\Gamma)$. It follows that $L(\Gamma) \cap \mathcal{D}_z(\Gamma)(\infty)$ is finite.

Consider now x in $L(\Gamma) \cap \mathcal{D}_z(\Gamma)(\infty)$. Since Γ is not elementary, the points of $L(\Gamma)$ are not isolated in $L(\Gamma)$. It follows that there is a non-stationary sequence $(x_n)_{n \geqslant 1}$ in $L(\Gamma)$ converging to x. One may assume that this sequence does not meet $\mathcal{D}_z(\Gamma)(\infty)$, and thus that there is some γ_1 in Γ satisfying: x is an endpoint of an edge $C(\gamma_1^{-1})$, and the sequence $(x_n)_{n \geqslant 1}$ is in the boundary at infinity of the half-plane $D(\gamma_1^{-1})$ (Fig. I.33).

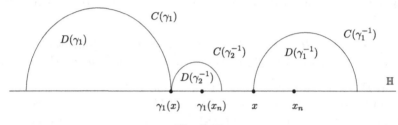

Fig. I.33.

The isometry γ_1 sends $C(\gamma_1^{-1})$ to the edge $C(\gamma_1)$, thus the point $\gamma_1(x)$ is an endpoint of $C(\gamma_1)$. Additionally, the sequence $(\gamma_1(x_n))_{n \geqslant 1}$ converges to $\gamma_1(x)$, is contained in $L(\Gamma)$, and never goes into $D(\gamma_1)(\infty)$. It follows that there is some γ_2^{-1} distinct from γ_1, such that $\gamma_1(x)$ is an endpoint of the edge $C(\gamma_2^{-1})$. Suppose that x is not parabolic. In this case, the isometry γ_2^{-1} is distinct from γ_1^{-1} by Exercise 2.7.

In the preceding argument, replace x with $\gamma_1(x)$ and γ_1 with γ_2. In so doing, if x is not parabolic one obtains an element γ_3 in $\Gamma - \{\gamma_2^{\pm 1}, \mathrm{Id}\}$ such that $\gamma_2\gamma_1(x)$ is in the boundary at infinity of $D(\gamma_2)$. Continuing this way, one constructs a sequence $(\gamma_n)_{n \geqslant 1}$ in $\Gamma - \{\mathrm{Id}\}$ satisfying

$$\gamma_n \cdots \gamma_1(x) \in \mathcal{D}_z(\Gamma)(\infty) \cap L(\Gamma) \quad \text{and} \quad \gamma_{n+1} \neq \gamma_n^{\pm 1}.$$

Since $\mathcal{D}_z(\Gamma)$ has finitely many edges, the set of points in this sequence is finite. Hence two integers $n < m$ can be chosen satisfying $\gamma_m \cdots \gamma_1(x) = \gamma_n \cdots \gamma_1(x)$. Because the point x is neither parabolic nor horocyclic, we have $\gamma_m \cdots \gamma_{n+1} = \mathrm{Id}$.

Let us examine the image of the point z, center of the Dirichlet domain $\mathcal{D}_z(\Gamma)$, by this element. The point $\gamma_{n+1}(z)$ is contained in the open half-plane $\mathring{D}(\gamma_{n+1})$. Furthermore, the half-planes $\mathring{D}(\gamma_{n+1})$ and $\mathring{D}(\gamma_{n+2})$ are disjoint. Hence $\gamma_{n+2}\gamma_{n+1}(z)$ is in $\mathring{D}(\gamma_{n+2})$. Continuing along these lines, one obtains that the point $\gamma_m \cdots \gamma_{n+1}(z)$ is in $\mathring{D}(\gamma_m)$, which contradicts the equality $\gamma_m \cdots \gamma_{n+1}(z) = z$. Thus we conclude that x is parabolic. Applying Corollary 4.10, we obtain that the set $L(\Gamma)$ is a finite union of Γ-orbits. □

The following theorem gives a characterization of the geometric finiteness in terms of points in the limit set.

Theorem 4.13. *Let Γ be a Fuchsian group. Then the following are equivalent:*

(i) *The group Γ is geometrically finite.*
(ii) $L(\Gamma) = L_p(\Gamma) \cup L_h(\Gamma)$.
(iii) $L(\Gamma) = L_p(\Gamma) \cup L_c(\Gamma)$.

Recall that the parabolic and conical points of the limit set of a Fuchsian group can be detected from the dynamics of this group on $\mathbb{H}(\infty)$ (Proposition 3.14). By Theorem 4.13, it turns out that the geometric finiteness of a Fuchsian group Γ, which was originally defined by the action of Γ on $\mathbb{H}$, is entirely determined by the dynamics of the group on $\mathbb{H}(\infty)$.

Notice that (iii) $\Rightarrow$ (ii) in Theorem 4.13 is clear since conical points are horocyclic.

Proof of (ii) $\Rightarrow$ (i) *in Theorem 4.13.*

Proposition 4.14. *If Γ is a Fuchsian group which is not geometrically finite, then for any Dirichlet domain $\mathcal{D}_z(\Gamma)$, there exists a point in $L(\Gamma) \cap \mathcal{D}_z(\Gamma)(\infty)$ which is not parabolic.*

Proof. Recall that $C(\gamma)$ denotes an edge of a Dirichlet domain $\mathcal{D}_z(\Gamma)$ which included in the perpendicular bisector of $[z, \gamma(z)]_h$.

Since Γ is not geometrically finite, any domain $\mathcal{D}_z(\Gamma)$ has infinitely many edges $(C(\gamma_n))_{n\geqslant 1}$. One can suppose that the sequence $(\gamma_n(z))_{n\geqslant 1}$ converges to some point x. Since the edge $C(\gamma_n)$ is included in the perpendicular bisectors $M_z(\gamma_n)$ of segments $[z, \gamma_n(z)]_h$, the sequence of edges $(C(\gamma_n))_{n\geqslant 1}$ converges to x and hence x is in $\mathcal{D}_z(\Gamma)(\infty) \cap L(\Gamma)$. One can suppose $x = \infty$. If x is parabolic, then the group Γ_x is generated by a non-trivial translation p. It follows that the domain $\mathcal{D}_z(\Gamma)$ is in the vertical strip bounded by the perpendicular bisectors of the segments $[z, g(z)]_h$ and $[z, g^{-1}(z)]_h$. On the other hand, the sequence of perpendicular bisectors $(M_z(\gamma_n))_{n\geqslant 1}$ converges to x and each $M_z(\gamma_n)$ meets $\mathcal{D}_z(\Gamma)$. Thus, for large enough n, the geodesic $M_z(\gamma_n)$ intersects $M_z(p)$ and $M_z(p^{-1})$, and the half-plane bounded by $M_z(\gamma_n)$ containing z is a half-disk perpendicular to the real axis. This contradicts the fact that the point ∞ is in $\mathcal{D}_z(\Gamma)(\infty)$. $\square$

We deduce from this proposition and from Corollary 4.10 the following corollary

Corollary 4.15 (Theorem 4.13(ii) $\Rightarrow$ (i)). *If $L(\Gamma) = L_p(\Gamma) \cup L_h(\Gamma)$, then Γ is geometrically finite.*

In this part, Γ is a non-elementary geometrically finite group and $\mathcal{D}_z(\Gamma)$ is a Dirichlet domain having finitely many edges.

Before giving the proof of (i) $\Rightarrow$ (iii) in Theorem 4.13, let us analyze the intersection of the Nielsen region $N(\Gamma)$ of the group Γ with $D_z(\Gamma)$.

First, we associate to each point x in the finite set $L_p(\Gamma) \cap \mathcal{D}_z(\Gamma)(\infty)$ a horodisk $H^+(x)$ centered at x satisfying the condition of Theorem 3.17

$$H^+(x) \cap \gamma H^+(x) = \varnothing,$$

for all γ in Γ not fixing x. One may choose the horodisks to be pairwise disjoint. Moreover one can assume that the horodisks

$$(\gamma(H^+(x)))_{\substack{\gamma \in \Gamma \\ x \in L_p(\Gamma) \cap \mathcal{D}_z(\Gamma)(\infty)}}$$

are pairwise either disjoint or identical.

To prove this, let x and y in $L_p(\Gamma) \cap \mathcal{D}_z(\Gamma)(\infty)$, with y not in $\Gamma(x)$. Consider the set A of $\gamma \in \Gamma$ such that $\gamma H^+(x) \cap H^+(y) \neq \varnothing$. Let B denote the set of two-sided cosets $\Gamma_y \backslash A / \Gamma_x$ (i.e., equivalence classes via two relations). If B is finite, it suffices to replace $H^+(x)$ with a smaller horodisk than $H^+(x)$, for which $\gamma H^+(x) \cap H^+(y) = \varnothing$ for all $\gamma \in \Gamma$. Otherwise, consider a sequence of distinct elements of B written as $(b_n = \Gamma_y a_n \Gamma_x)_{n \geqslant 1}$, where $a_n \in A$. Fix a compact fundamental domain K for the action of Γ_y on $H(y)$. For all $n \geqslant 1$, there exists $p_n \in \Gamma_y$ such that $p_n a_n H(x) \cap K \neq \varnothing$. The horocycles $(p_n a_n H(x))_{n \geqslant 1}$ are pairwise disjoint but also intersect K, which is impossible.

The following proposition gives a decomposition of the intersection of the Nielsen region $N(\Gamma)$ of the group Γ with $\mathcal{D}_z(\Gamma)$.

Proposition 4.16. *Let Γ be a non-elementary geometrically finite group. Then there exists a relatively compact set $K \subset \mathbb{H}$ such that*

$$N(\Gamma) \cap \mathcal{D}_z(\Gamma) = K \bigcup_{x \in L_p(\Gamma) \cap \mathcal{D}_z(\Gamma)(\infty)} H^+(x) \cap \mathcal{D}_z(\Gamma).$$

Proof. Suppose that the closure of the intersection of the Nielsen region $N(\Gamma)$ with the set $\mathcal{D}_z(\Gamma) - \bigcup_{x \in L_p(\Gamma) \cap \mathcal{D}_z(\Gamma)(\infty)} H^+(x) \cap \mathcal{D}_z(\Gamma)$ is not compact. Consider y in the intersection of this closure with $\mathbb{H}(\infty)$. Such a point is not in $L(\Gamma)$. Otherwise, y is in $\mathcal{D}_z(\Gamma)(\infty) \cap L(\Gamma)$ and hence, since $\mathcal{D}_z(\Gamma)$ has finitely many edges, y would be a parabolic point (Theorem 4.12). Let p be a generator of Γ_y. The domain $\mathcal{D}_z(\Gamma)$ is included in the intersection of the half-planes

$$\mathbb{H}(p) = \{z' \in \mathbb{H} \mid d(z', z) \leqslant d(z', p(z))\} \quad \text{and}$$
$$\mathbb{H}(p^{-1}) = \{z' \in \mathbb{H} \mid d(z', z) \leqslant d(z', p^{-1}(z))\}.$$

It follows that set $\mathcal{D}_z(\Gamma) - \bigcup_{x \in L_p(\Gamma) \cap \mathcal{D}_z(\Gamma)(\infty)} H^+(x) \cap \mathcal{D}_z(\Gamma)$ is included in the set $\mathbb{H}(p) \cap \mathbb{H}(p^{-1}) - \mathbb{H}(p) \cap \mathbb{H}(p^{-1}) \cap H^+(y)$, which is impossible since the boundary at infinity of this last set does not contain y.

In conclusion, the point y is not in $L(\Gamma)$ and hence is in an open interval I in $\mathcal{D}_z(\Gamma)(\infty)$ which does not intersect $L(\Gamma)$. Let L be the geodesic whose endpoints are the endpoints of I. The set $\hat{\Omega}(\Gamma)$, and hence $N(\Gamma)$, is in the closed half-plane bounded by L, whose boundary at infinity is $(\mathbb{H}(\infty) - I)$. Since y is in the boundary at infinity of $N(\Gamma)$, this point is in $(\mathbb{H}(\infty) - I) \cap I$, which is not possible. In conclusion, the intersection of the Nielsen region $N(\Gamma)$ with the set $\mathcal{D}_z(\Gamma) - \bigcup_{x \in L_p(\Gamma) \cap \mathcal{D}_z(\Gamma)(\infty)} H^+(x) \cap \mathcal{D}_z(\Gamma)$ is relatively compact in $\mathbb{H}$. $\square$

Proof of (i) $\Rightarrow$ (iii) *in Theorem 4.13.*

Suppose that Γ is geometrically finite. We are ready now to prove that $L(\Gamma) = L_p(\Gamma) \cup L_c(\Gamma)$. Since Γ is not elementary, the set $L_c(\Gamma)$ is not empty. Take any point y in $L(\Gamma) - L_p(\Gamma)$, and let z' in $\mathbb{H}$ be such that $[z', y)$ is in $N(\Gamma)$. Since $\mathcal{D}_z(\Gamma)$ is a fundamental domain, the ray $[z', y)$ is contained in the union $\bigcup_{\gamma \in \Gamma} \gamma(\mathcal{D}_z(\Gamma)) \cap [z', y)$. Suppose that there is z'' in $[z', y)$ such that $[z'', y)$ is contained in $\bigcup_{\gamma \in \Gamma, x \in L_p(\Gamma) \cap \mathcal{D}_z(\Gamma)(\infty)} \gamma(H^+(x))$. Since the horodisks $(\gamma(H^+(x)))_{\gamma \in \Gamma, x \in L_p(\Gamma) \cap \mathcal{D}_z(\Gamma)(\infty)}$ are either disjoint or identical, the ray $[z'', y)$ is contained in one such horodisk, which is impossible since y is not parabolic.

From Proposition 4.16, we obtain a compact $K \subset \mathbb{H}$, a sequence $(z_n)_{n \geqslant 1}$ in $[z', y)$ converging to y, and $(\gamma_n)_{n \geqslant 1}$ in Γ, such that z_n is in $\gamma_n(K)$. Since K is compact, the sequence $(\gamma_n(z'))_{n \geqslant 1}$ converges to y while remaining in an ε-neighborhood of $[z', y)$. This shows that y is conical.

Recall that a geometrically finite Fuchsian Γ group is said to be convex-cocompact if for some Dirichlet domain $D_z(\Gamma)$, the set $D_z(\Gamma) \cap N(\Gamma)$ is compact. From Theorems 4.12 and 4.13, we deduce the following characterization of convex-cocompact groups and of lattices in terms of their points at infinity.

Corollary 4.17. *Let Γ be a Fuchsian group.*

- *The group Γ is convex-cocompact if and only if $L(\Gamma) = L_c(\Gamma)$.*
- *The group Γ is a lattice if and only if $\mathbb{H}(\infty) = L_p(\Gamma) \cup L_c(\Gamma)$.*

5 Comments

The notions and main results of this chapter can be generalized to the case of an oriented Riemannian manifold X called a *pinched Hadamard manifold*, of dimension $n \geqslant 2$, which is simply connected, complete and has sectional curvature which is bounded by two strictly negative constants [12, 6, 14, 28]. Below we give a broad outline of this generalization.

In this context, *geodesics* are well-defined and the boundary at infinity $X(\infty)$ of this manifold is defined to be the set of equivalence classes of asymptotic geodesic rays. While viewing the points of X as endpoints of geodesic segments of fixed origin, one can provide $X \cup X(\infty)$ with a natural topology, which in fact is a compactification of X in which X is an open dense subset. For this compactification, the set of oriented geodesics of X is identified with the pairs of distinct points of $X(\infty)$. The notion of *Busemann cocycles* (Sect. 1.4) can also be extended to this general setting and allows us to define *horocycles* (for $n = 2$) and *horospheres* (for $n \geqslant 3$) of X.

On the other hand, unlike the transitivity of the action of G on $\mathbb{H}$, the group of positive isometries of X may be very poor, possibly even trivial. The proofs given in this chapter which require transitivity cannot be directly adapted to the general case. However, the majority of them can be translated into Riemannian terms. This is the case, for example, in the proof of Property 1.19.

As with $\mathbb{H}$, if a positive isometry does not fix any point in X, then it fixes exactly one (*parabolic* isometry) or two (*hyperbolic* isometry) points in $X(\infty)$.

We call a subgroup of positive isometries whose action is properly discontinuous on X a *Kleinian group*. The existence of such non-trivial groups is not guaranteed.

When such a group does exist, the definitions of *limit set*, *horocyclic* (*horospheric*), *conical* and *parabolic* points, given in Sect. 3, remain unchanged. In our text, we concentrate on a category of Fuchsian groups for which conical and horocyclic points are interchangeable. In an article by A. Starkov [59], the reader will find some examples of groups having horocyclic points which are not conical.

The notion of geometric finiteness has arisen in the context of hyperbolic space of curvature -1 in dimension 3, the motivation being to study the action of discrete groups of finite type on this space. When $\dim X = 2$, the notion of discrete groups of finite type and geometrically finite groups coincide. This is not the case in higher dimensions [10].

When the dimension of X is $\geqslant 3$, the stabilizer of a parabolic point x in Γ does not necessarily act cocompactly on $L(\Gamma) - \{x\}$. If this action is co-compact, x is called a *bounded parabolic point*. This family of points emerges in an essential way in the generalization of the notion of geometric finiteness of a group to pinched Hadamard manifolds. In this setting, the definition is delicate and can be formulated in several ways [12]. One of these formulations rests on the decomposition of $L(\Gamma)$ into conical and bounded parabolic points. Under this hypothesis, the number of Γ-orbits of parabolic points is finite. Each of these points x has an associated horodisk (horoball) $H^+(x)$ whose image by the group Γ is either disjoint from or identical to $H^+(x)$ [54, Lemma 1.9]. The quotient of $H^+(x)$ by the stabilizer of x in Γ injects into the manifold $M = \Gamma \backslash X$, producing what we will continue to call a *cusp*. If Γ is a geometrically finite Kleinian group, the set of points of X contained in geodesics whose endpoints are in the limit set of the group, projects into the union of a compact subset of M and a finite number of cusps. In this context it is relatively simple to see that most of the results about the topological dynamics of geodesic flow and horospheric foliations proved in the following chapters can be generalized.

We conclude these comments with an outline of some metric properties of the limit set of a non-elementary Fuchsian group acting on the Poincaré disk $\mathbb{D}$.

One of the keys to this rich area requires the development of the *Poincaré series* $P_s(\Gamma)$ of Γ which is defined by

$$P_s(\Gamma) = \sum_{\gamma \in \Gamma} e^{-sd(0,\gamma(0))},$$

where 0 is the center of the disk $\mathbb{D}$. Its critical exponent $\delta(\Gamma)$ can also be defined by [54]:

$$\delta(\Gamma) = \lim_{R \to +\infty} \frac{1}{R}(\ln \operatorname{card}\{\gamma \in \Gamma \mid d(0, \gamma(0)) \leqslant R\}).$$

This series allows us to establish a relationship between the statistical behavior of $\Gamma(0)$ and the metric properties of $L(\Gamma)$. One can show for example that if Γ is geometrically finite, then $\delta(\Gamma)$ is equal to the Hausdorff dimension of $L(\Gamma)$ ([51], [48, Theorem 9.3.6]).

This series also allows us to construct measures m, called *Patterson measures*, whose support is $L(\Gamma)$ and which, while not being Γ-invariant (Γ-invariant measures do not exist if Γ is not elementary), are conformal in the sense that they satisfy the relation

$$\forall \gamma \in \Gamma, \quad \frac{d\gamma^{-1}m}{dm}(x) = |\gamma'(x)|^{\delta(\Gamma)},$$

where $|\gamma'(x)|$ represents the conformal factor at the point x of the map γ, seen as a conformal transformation [51, 48] of the disk $\mathbb{D}$.

The construction of these measures is due to S. Patterson ([51], [8, Chap. 9]). If the series $P_s(\Gamma)$ diverges for $s = \delta(\Gamma)$, which is the case when Γ is geometrically finite [48], such a measure is obtained by taking the weak limit, when s tends to $\delta(\Gamma)$ from above, of a sequence of orbital measures m^s defined by

$$m^s = \frac{1}{P_s(\Gamma)} \sum_{\gamma \in \Gamma} e^{-sd(0,\gamma(0))} D_{\gamma(0)},$$

where $D_{\gamma(0)}$ represents the Dirac (point mass) measure at $\gamma(0)$.

If Γ is a lattice, this measure is proportional to the Lebesgue measure on $\mathbb{D}(\infty)$. As we will show in the Comments following Chaps. III and V, one interesting aspect of Patterson measures is that they allow a construction of measures on $\Gamma \backslash T^1 \mathbb{H}$ which is invariant with respect to geodesic flow and horocyclic foliation.

The construction of Patterson measures on $L(\Gamma)$ can be generalized to Kleinian groups acting on pinched Hadamard manifolds [11, 54].

II

Examples of Fuchsian groups

In this chapter, we study concrete examples of Fuchsian groups and illustrate the results of the previous chapter.

The first family of groups that we will consider consists of geometrically finite free groups, called *Schottky* groups. Its construction is based on the dynamics of isometries.

The second family comes from number theory. It consists of three non-uniform lattices: the *modular* group $\mathrm{PSL}(2,\mathbb{Z})$, its congruence modulo 2 subgroup and its commutator subgroup.

We will study each of these groups according the same general outline:

- description of a fundamental domain;
- shape of the associated topological surface;
- properties of its isometries;
- study of its limit set;
- characterization of its parabolic points.

We will also construct a coding of the limit sets of Schottky groups and of the modular group. We will use this coding in Chap. IV to study the dynamics of the geodesic flow, and in Chap. VII to translate the behavior of geodesic rays on the modular surface into the terms of Diophantine approximations.

1 Schottky groups

The Poincaré disk model $\mathbb{D}$ is the ambient space in this discussion. We will fix a point 0 in this set, which is not necessary the origin of the disk.

Recall that, if g is a positive isometry of $\mathbb{D}$ which does not fix 0, then the set $D_0(g) = D(g)$ represents the closed half-plane in $\mathbb{D}$ bounded by the perpendicular bisector of the hyperbolic segment $[0, g(0)]_h$, containing $g(0)$. The sets $D(g)$ and $D(g^{-1})$ are disjoint (resp. tangent) if and only if g is hyperbolic (resp. parabolic) (Property I.2.7). Moreover we have:

$$g(D(g^{-1})) = \mathbb{D} - \overset{\circ}{D}(g).$$

F. Dal'Bo, *Geodesic and Horocyclic Trajectories*, Universitext,
DOI 10.1007/978-0-85729-073-1_2, © Springer-Verlag London Limited 2011

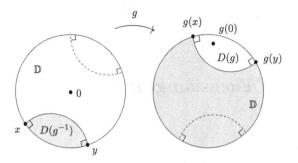

Fig. II.1. g hyperbolic

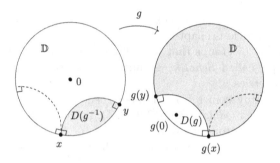

Fig. II.2. g parabolic

Definition 1.1. Let p be an integer $\geqslant 2$. A Schottky group of rank p is a subgroup of G which has a collection of non-elliptic, non-trivial generators $\{g_1, \ldots, g_p\}$ satisfying the following condition: there exists a point 0 in $\mathbb{D}$ such that the closures in $\overline{\mathbb{D}} = \mathbb{D} \cup \mathbb{D}(\infty)$ of the sets $D_0(g_i^{\pm 1}) = D(g_i^{\pm 1})$, for $i = 1, \ldots, p$, satisfy

$$\overline{\left(D(g_i) \cup D(g_i^{-1})\right)} \cap \overline{\left(D(g_j) \cup D(g_j^{-1})\right)} = \varnothing,$$

for all $i \neq j$ in $\{1, \ldots, p\}$.

Let $S(g_1, \ldots, g_p)$ denote such a group whose collection of generators is $\{g_1, \ldots, g_p\}$.

In the rest of this discussion, in order to avoid notational clutter, we will restrict ourselves to the case where $p = 2$.

Figure II.3 represents the four possible configurations associated with Schottky groups of rank 2.

Schottky groups are not especially difficult to find. The following lemma shows that their construction only requires two non-elliptic isometries.

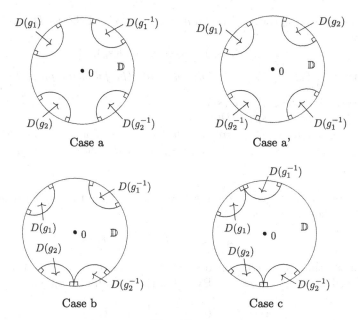

Fig. II.3.

Lemma 1.2. *Let g and g' be two non-elliptic isometries in G which have no common fixed points. Then there exists $N > 0$ such that g^N and g'^N generate a Schottky group $S(g^N, g'^N)$.*

Proof. Fix a point 0 in $\mathbb{D}$. The sequence $(g^n(0))_{n \geqslant 1}$ converges to a point x which is fixed by g. Thus the sequence of perpendicular bisectors of $[0, g^n(0)]_h$ similarly converges to x. Since g and g' do not have any fixed points in common, for large enough n the closed sets $\overline{D(g^n) \cup D(g^{-n})}$ and $\overline{D(g'^n) \cup D(g'^{-n})}$ are disjoint. □

Since a non-elementary Fuchsian group contains infinitely many non-elliptic isometries which have no common fixed points, we deduce from Lemma 1.2 the following result

Corollary 1.3. *A non-elementary Fuchsian group contains infinitely many Schottky groups.*

1.1 Dynamics of Schottky groups

Fix a Schottky group $S(g_1, g_2)$ of rank 2. The *alphabet* of this group is by definition the set $\mathcal{A} = \{g_1^{\pm 1}, g_2^{\pm 1}\}$. A product of n letters $s_1 \cdots s_n$ in $\mathcal{A}$ is said to be a *reduced word* of $S(g_1, g_2)$ if $n = 1$, or if $n > 1$ and $s_i \neq s_{i+1}^{-1}$ for all $1 \leqslant i \leqslant n - 1$. The integer n is called the *length* of $s_1 \cdots s_n$. We associate to a reduced word $s_1 \cdots s_n$ the set $D(s_1)$ if $n = 1$, and if $n > 1$ the set $D(s_1, \ldots, s_n)$ defined by:

$$D(s_1, \ldots, s_n) = s_1 \cdots s_{n-1} D(s_n).$$

Property 1.4. *Let $s_1 \cdots s_n$ be a reduced word. The following properties hold:*

(i) $s_1 \cdots s_n (\mathbb{D} - \mathring{D}(s_n^{-1})) \subset D(s_1)$.
(ii) *If* $n \geqslant 2$, $D(s_1, \ldots, s_n) \subset D(s_1, \ldots, s_{n-1})$.
(iii) *If $s_1 \cdots s_n$ and $s_1' \cdots s_n'$ are two distinct reduced words, then the half-planes $D(s_1, \ldots, s_n)$ and $D(s_1', \ldots, s_n')$ are either tangent or disjoint.*

Proof.

(i) We prove (i) by induction on $n \geqslant 1$. When $n = 1$, part (i) is a consequence of the following relation:

$$\forall s \in \mathcal{A}, \quad s(\mathbb{D} - \mathring{D}(s^{-1})) = D(s).$$

Suppose that (i) is true up to $n = N$, for some $N \geqslant 1$. Consider the reduced word $s_1 \cdots s_N s_{N+1}$. By induction hypothesis, one has

$$s_2 \cdots s_{N+1} (\mathbb{D} - \mathring{D}(s_{N+1}^{-1})) \subset D(s_2).$$

Since $s_2 \neq s_1^{-1}$, the set $D(s_2)$ is contained in $\mathbb{D} - \mathring{D}(s_1^{-1})$. Furthermore, the set $s_1(\mathbb{D} - \mathring{D}(s_1^{-1}))$ is equal to $D(s_1)$, thus

$$s_1 \cdots s_{N+1} (\mathbb{D} - \mathring{D}(s_{N+1}^{-1})) \subset D(s_1).$$

(ii) Since $s_n \neq s_{n-1}^{-1}$, one has $D(s_n) \subset \mathbb{D} - \mathring{D}(s_{n-1}^{-1})$. Thus the set $s_{n-1} D(s_n)$ is contained in $D(s_{n-1})$, which implies (ii).
(iii) Let $k \geqslant 1$ be the smallest integer $\leqslant n$ such that $s_k' \neq s_k$. Proving part (iii) reduces to proving that $D(s_k', \ldots, s_n')$ and $D(s_k, \ldots, s_n)$ are tangent or disjoint.
By (i) and (ii), the set $D(s_k', \ldots, s_n') = s_k' \cdots s_{n-1}' D(s_n')$ is a subset of $D(s_k')$. Likewise $D(s_k, \ldots, s_n)$ is a subset of $D(s_k)$. Since $s_k \neq s_k'$, the sets $D(s_k)$ and $D(s_k')$ are tangent or disjoint, hence the half-planes $D(s_k', \ldots, s_n')$ and $D(s_k, \ldots, s_n)$ are as well. □

It is of interest to note that the only hypothesis which played a role in proving these properties was the following: if a and b are contained in $\mathcal{A}$ with $a \neq b^{-1}$, then $D(a)$ is contained in $\mathbb{D} - \mathring{D}(b^{-1})$. As a result, these properties remain valid for *generalized Schottky* groups of rank p. This means that the group admits a collection of non-elliptic generators $\{g_1, \ldots, g_p\}$ satisfying the following weaker condition:

$$(D(g_i) \cup D(g_i^{-1})) \cap (D(g_j) \cup D(g_j^{-1})) = \varnothing,$$

for all $i \neq j$ in $\{1, \ldots, p\}$.
We will meet such groups again in Sect. 3.

Exercise 1.5. Prove that in each of the cases a, a', b, c represented in Fig. II.3, the configuration of half-planes $D(g_1, g_2)$ is as follows in Fig. II.4.

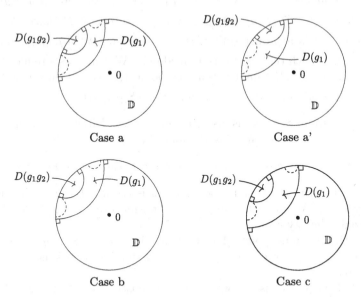

Fig. II.4.

Let g in G and $z \in \mathbb{D}$ such that $g(z) \neq z$. Following the notation used in Chap. I, we denote by $\mathbb{D}_z(g)$ the closed half-plane containing z, bounded by the perpendicular bisector of the segment $[z, g(z)]_h$. We have: $\mathbb{D}_z(g) = \mathbb{D} - \mathring{D}_z(g)$. Recall that the Dirichlet domain $\mathcal{D}_z(\Gamma)$ of a Fuchsian group Γ centered at z is the intersection of all sets $\mathbb{D}_z(\gamma)$, with γ in $\Gamma - \{\text{Id}\}$.

Proposition 1.6. *The group $S(g_1, g_2)$ is free with respect to g_1, g_2, and is discrete. Furthermore, the set $\bigcap_{\substack{i=1,2 \\ \varepsilon=\pm 1}} \mathbb{D} - \mathring{D}(g_i^\varepsilon)$ is the Dirichlet domain of this group centered at the point 0.*

Proof. Let $s_1, \ldots, s_n$ be a reduced word. From Property 1.4(i), the point $s_1 \cdots s_n(0)$ is contained is $D(s_1)$. The sets $D(s_1)$ and $\mathring{\mathcal{D}}(S(g_1, g_2))$ are disjoint. In particular $s_1 \cdots s_n(0) \neq 0$, which shows that $S(g_1, g_2)$ is free.

Let us prove that $S(g_1, g_2)$ is discrete. Consider a sequence $(\gamma_n)_{n \geq 1}$ in $S(g_1, g_2) - \{\text{Id}\}$. Each γ_n can be written as a reduced word $s_{n,1} \cdots s_{n,\ell_n}$. Passing to a subsequence $(\gamma_{n_k})_{k \geq 1}$, one may assume that $s_{n_k,1} = s_1$ for all $k \geq 1$. The point $\gamma_{n_k}(0)$ is contained in $D(s_1)$, thus there exists $c > 0$ such that $d(\gamma_{n_k}(0), 0) > c$ for all $k \geq 1$. This shows that $(\gamma_n)_{n \geq 1}$ cannot converge to the identity and thus $S(g_1, g_2)$ is discrete.

Since $S(g_1, g_2)$ is free and discrete, none of its elements is elliptic. Therefore, the Dirichlet domain $\mathcal{D}_0(S(g_1, g_2))$ centered at 0 is well defined.

It only remains to show that $F = \bigcap_{\substack{i=1,2 \\ \varepsilon=\pm 1}} \mathbb{D} - \overset{\circ}{D}(g_i^\varepsilon)$ and $\mathcal{D}_0(S(g_1, g_2))$ are equal. The set $\mathcal{D}_0(S(g_1, g_2))$ is certainly a subset of F, since $\mathbb{D}_0(g_i) = \mathbb{D} - \overset{\circ}{D}(g_i)$. If it is a proper subset, there exist z in $\mathcal{D}_0(S(g_1, g_2))$ and γ in $S(g_1, g_2) - \{\mathrm{Id}\}$ such that $\gamma(z)$ is contained in $\overset{\circ}{F}$. Writing γ as a reduced word $s_1 \cdots s_n$, Property 1.4(i) implies that the point $\gamma(z)$ is an element of $D(s_1)$. This is impossible since $\gamma(z)$ is contained in $\overset{\circ}{\mathcal{D}}(S(g_1, g_2))$.   □

Since the group $S(g_1, g_2)$ is discrete and admits a Dirichlet domain having finitely many edges, one can state the following result:

Corollary 1.7. *The group $S(g_1, g_2)$ is a geometrically finite Fuchsian group.*

Notice that the proof of Proposition 1.6 is essentially an application of Property 1.4(i). As such, this proposition and its corollary are also valid for generalized Schottky groups.

From the dynamic point of view, Schottky groups $S(g_1, g_2)$, where g_1 and g_2 are hyperbolic, are—in some sense—the simplest non-elementary Fuchsian groups.

The following exercise shows that the construction of these Schottky groups can be extended to an infinite collection of generators.

Exercise 1.8. Let $(g_i)_{i \geqslant 1}$ be an infinite sequence of non-elliptic isometries in G satisfying the following condition for all $i \neq j$:

$$(D(g_i) \cup D(g_i^{-1})) \cap (D(g_j) \cup D(g_j^{-1})) = \varnothing.$$

Prove that the group generated by this sequence is a Fuchsian free group which is not geometrically finite.

We now focus on the nature of the isometries of the $S(g_1, g_2)$.

Property 1.9. *Let $S(g_1, g_2)$ be a Schottky group.*

(i) *If g_1 and g_2 are both hyperbolic, then every element of $S(g_1, g_2) - \{\mathrm{Id}\}$ is hyperbolic.*

(ii) *If not, the non-hyperbolic isometries in $S(g_1, g_2) - \{\mathrm{Id}\}$ are conjugate to powers of parabolic generators in $S(g_1, g_2)$.*

Proof. Since the group $S(g_1, g_2)$ is free, it does not contain elliptic isometries. Let x be a point in $\mathbb{D}(\infty)$ fixed by a non-trivial parabolic isometry of $S(g_1, g_2)$. Applying Corollary I.4.10, we obtain γ in Γ such that $y = \gamma(x)$ is in $\mathcal{D}_0(S(g_1, g_2))(\infty)$. Since y is in $L(S(g_1, g_2))$, this point is an endpoint of an edge of $\mathcal{D}_0(S(g_1, g_2))$. Using the dynamics of the parabolic isometries, we have that any open arc in $\mathbb{D}(\infty)$ with extremity y meets $L(S(g_1, g_2))$. It follows that y is the common endpoint of two edges, and hence that it is fixed

by a parabolic generator $a \in \mathcal{A}$ (see Fig. II.3). The group $S(g_1, g_2)$ being discrete, any isometry in $S(g_1, g_2)$ fixing y belongs to the cyclic group generated
by a. This implies that x is fixed by an isometry of the form $\gamma a^k \gamma^{-1}$, for some
$k \neq 0$.   □

These properties cannot be extended to generalized Schottky groups. It
will be shown in the next section that some such groups can contain parabolic
isometries which are not conjugate to powers of generators.
Using Proposition I.2.17, we obtain that the surfaces $S(g_1, g_2) \backslash \mathbb{D}$ associated to each of the four cases in Fig. II.3 are of the form shown in Fig. II.5.

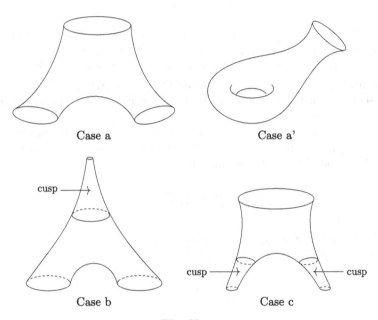

Fig. II.5.

1.2 Limit set

Clearly, the limit set of a Schottky group $S(g_1, g_2)$ is a proper subset of
$\mathbb{D}(\infty)$ since it does not meet the non-empty, open, circular arcs included in
$\mathcal{D}_0(S(g_1, g_2))(\infty)$.
What is the topological structure of $L(S(g_1, g_2))$? To begin answering this
question, we prove the following lemma:

Lemma 1.10. *Given a sequence* $(s_i)_{i \geq 1}$ *in* $\mathcal{A}$ *satisfying* $s_{i+1} \neq s_i^{-1}$, *the sequence of Euclidean diameters of the sets* $D(s_1, \ldots, s_n)$ *converges to zero.*

Proof. By Property 1.4(ii), the half-planes $D(s_1, \ldots, s_n)$ are nested. If the sequence of Euclidean radii do not converge to 0, there is a compact set K in $\mathbb{D}$ such that for all geodesics of the form $s_1 \cdots s_{n-1}(\mathcal{C}_n)$, where $\mathcal{C}_n$ is the boundary of $D(s_n)$, we have $K \cap s_1 \cdots s_{n-1}(\mathcal{C}_n) \neq \emptyset$. The geodesics $\mathcal{C}_n$ are edges of the Dirichlet domain $\mathcal{D}_0(S(g_1, g_2))$, thus the image of this domain under the maps $s_1 \cdots s_{n-1}$ intersects K. Yet this is impossible since Property I.2.15 states that this domain is locally finite.   □

Proposition 1.11.

$$L(S(g_1, g_2)) = \bigcap_{n=1}^{+\infty} \bigcup_{\substack{\text{reduced words} \\ \text{of length } n}} \overline{D(s_1, \ldots, s_n)}.$$

Proof. Let y be an element of $L(S(g_1, g_2))$, and consider a sequence $(\gamma_n)_{n \geqslant 1}$ in $S(g_1, g_2)$ such that $\lim_{n \to +\infty} \gamma_n(0) = y$. Write γ_n as a reduced word $\gamma_n = s_{n,1} \cdots s_{n,\ell_n}$, where $s_{n,i} \in \mathcal{A}$ and $s_{n,i} \neq s_{n,i+1}^{-1}$. Since $\mathcal{A}$ is finite, one can assume (by passing to a subsequence) that there exist $(s_i)_{i \geqslant 1}$ with $s_{i+1} \neq s_i^{-1}$, and a sequence of positive integers $(\ell_n)_{n \geqslant 1}$ which is strictly increasing such that $\gamma_n = s_1 \cdots s_{\ell_n}$.

The point $s_{\ell_n}(0)$ is an element of $D(s_{\ell_n})$, therefore $\gamma_n(0)$ is in $D(s_1, \ldots, s_{\ell_n})$, for any $n \geqslant 1$. Since the sets $D(s_1, \ldots, s_n)$ are nested and their diameter go to 0 (Lemma 1.10), we have

$$\{y\} = \bigcap_{n=1}^{+\infty} \overline{D(s_1, \ldots, s_{\ell_n})}.$$

This shows that $L(S(g_1, g_2))$ is a subset of

$$\bigcap_{n=1}^{+\infty} \bigcup_{\substack{\text{reduced words} \\ \text{of length } n}} \overline{D(s_1, \ldots, s_n)}.$$

The reverse inclusion is a consequence of Property 1.4 and Lemma 1.10.   □

Using this proposition, we obtain a construction of $L(S(g_1, g_2))$ by an iterative procedure analogous to the construction of *Cantor sets*. More precisely, consider the case in which g_1 and g_2 are hyperbolic and denote by I_1, I_2, I_3, I_4 the connected components of the following set:

$$\mathbb{D}(\infty) - \bigcup_{a \in \mathcal{A}} D(a)(\infty).$$

Step 1: Remove these four arcs from the set $\mathbb{D}(\infty)$. One obtains

$$\mathbb{D}(\infty) - \bigcup_{1 \leqslant i \leqslant 4} I_i = \bigcup_{a \in \mathcal{A}} D(a)(\infty).$$

Step 2: For each a, remove the four arcs $a(I_1), a(I_2), a(I_3), a(I_4)$ from the set $D(a)(\infty)$. One obtains

$$\forall\, a \in \mathcal{A}, \quad \mathbb{D}(a)(\infty) - \bigcup_{1 \leqslant i \leqslant 4} (I_i) = \bigcup_{1 \leqslant i \leqslant 4} \bigcup_{\substack{b \in \mathcal{A} \\ b \neq a^{-1}}} D(a,b)(\infty).$$

More generally, one does the following for $n \geqslant 2$:
Step n: Remove from each set of the form $D(s_1, \dots, s_{n-1})(\infty)$, where $s_1 \cdots s_{n-1}$ is a reduced word, the four arcs $s_1 \cdots s_{n-1}(I_1)$, $s_1 \cdots s_{n-1}(I_2)$, $s_1 \cdots s_{n-1}(I_3)$, $s_1 \cdots s_{n-1}(I_4)$. One obtains

$$D(s_1, \dots, s_{n-1})(\infty) - \bigcup_{1 \leqslant i \leqslant 4} s_1 \cdots s_{n-1}(I_i) = \bigcup_{\substack{s \in \mathcal{A} \\ s \neq s_{n-1}^{-1}}} D(s_1, \dots, s_{n-1}, s)(\infty).$$

It follows from Proposition 1.11, that the set $L(S(g_1, g_2))$ is the intersection over the integers $n \geqslant 1$, of $4 \times 3^{n-1}$ arcs obtained at Step n of this procedure (Fig. II.6).

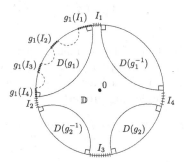

Fig. II.6.

If some of the generators is parabolic, then the procedure above must be modified by grouping together arcs of the form $D(s_1, \dots, s_n)(\infty)$ and $D(s_1', \dots, s_n')(\infty)$ having an endpoint in common.

Exercise 1.12. Prove that the set $L(S(g_1, g_2))$ is a totally discontinuous set (i.e., its connected components are points), without isolated points. (Hint: [13].)

Recall that the Nielsen region $N(S(g_1, g_2))$ of $S(g_1, g_2)$ is the convex hull of the set of points in $\mathbb{D}$ belonging to geodesics whose endpoints are in the limit set $L(S(g_1, g_2))$ (Sect. I.4).

Figure II.7 shows the form of the intersection of $N(S(g_1, g_2))$ with the Dirichlet domain $\mathcal{D}_0(S(g_1, g_2))$ in cases a, a', b, c associated with Fig. II.3.

If neither of its generators are parabolic, the group $S(g_1, g_2)$ is geometrically finite and does not contain any parabolic isometries, and thus this group

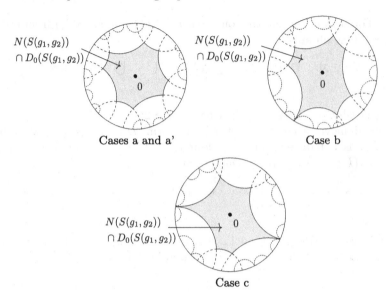

$$N(S(g_1,g_2))$$
$$\cap D_0(S(g_1,g_2))$$

$$N(S(g_1,g_2))$$
$$\cap D_0(S(g_1,g_2))$$

Cases a and a'   Case b

$$N(S(g_1,g_2))$$
$$\cap D_0(S(g_1,g_2))$$

Case c

Fig. II.7.

is convex-cocompact (Corollary I.4.17). This property is shown by Fig. II.7 (cases a and a') since the set $N(S(g_1,g_2)) \cap D_0(S(g_1,g_2))$ is compact (Definition I.4.5). This is not the case if at least one of the generators is parabolic.

Thus one can state the following property.

Property 1.13. *The group* $S(g_1,g_2)$ *is convex-cocompact if and only if* g_1 *and* g_2 *are hyperbolic.*

2 Encoding the limit set of a Schottky group

At this stage, we enter the world of symbolic dynamics which will be explored further in Chap. IV.

The purpose of this section is simply to construct a dictionary between $L(S(g_1,g_2))$ and a specific set of sequences of elements of $\mathcal{A}$. Using this dictionary, we establish a correspondence between some properties of these sequences and some geometric properties of points in $L(S(g_1,g_2))$.

Since the group $S(g_1,g_2)$ is free, there is a bijection between the set of finite reduced sequences $s_1,\ldots,s_n$, with $s_i \neq s_{i+1}^{-1}$, and $S(g_1,g_2) - \{\mathrm{Id}\}$. Let us extend this bijection to the set of infinite reduced sequences Σ^+ defined by

$$\Sigma^+ = \{(s_i)_{i \geq 1} \mid s_i \in \mathcal{A}, \ s_{i+1} \neq s_i^{-1}\}.$$

Let $(s_i)_{i \geq 1}$ be such a sequence. Define

$$\gamma_n = s_1 \cdots s_n.$$

The point $\gamma_n(0)$ is in the half-plane $D(s_1, \ldots, s_n)$. As a consequence of Property 1.4(ii), the half-planes $(D(s_1, \ldots, s_n))_{n \geqslant 1}$ are nested and, by Lemma 1.10, their Euclidean diameter tends to 0. Hence the sequence $(\gamma_n(0))_{n \geqslant 1}$ converges to a point in $L(S(g_1, g_2))$.

Let $f : \Sigma^+ \to L(S(g_1, g_2))$ denote the function which sends $s = (s_i)_{i \geqslant 1}$ to the point x defined by

$$x(s) = \lim_{n \to +\infty} \gamma_n(0).$$

Exercise 2.1. Prove that the function f is surjective.
(Hint: Use Property 1.11.)

Is this function injective? To answer this question, the cases with and without parabolic generators must be considered separately.

Consider the subset Σ_c^+ of Σ^+ consisting of sequences $(s_n)_{n \geqslant 1} \in \Sigma^+$ for which, if the term s_n is parabolic, there exists $m > n$ such that $s_m \neq s_n$.

Recall that, since $S(g_1, g_2)$ is geometrically finite, its limit set can be decomposed into a disjoint union of the set of its parabolic points $L_p((S(g_1, g_2))$ and its conical points $L_c(S(g_1, g_2))$ (Theorem I.4.13).

Proposition 2.2. If g_1 and g_2 are hyperbolic, then the function $f : \Sigma^+ \to L(S(g_1, g_2))$ is a bijection. Otherwise, this function is surjective but not injective and its restriction to Σ_c^+ is a bijection onto $L_c(S(g_1, g_2))$.

Proof.
Case 1: g_1 and g_2 hyperbolic. In this case $\Sigma_c^+ = \Sigma^+$ and for all a in $\mathcal{A}$ and b in $\mathcal{A} - \{a\}$, the closure of the sets $D(a)$ and $D(b)$ are disjoint. Let $s = (s_i)_{i \geqslant 1}$ and $s' = (s'_i)_{i \geqslant 1}$ be in Σ^+. Suppose that there exists $i \geqslant 1$ such that $s_i \neq s'_i$. Let k denote the smallest of these integers and define $\gamma = s_1 \cdots s_{k-1}$ if $k > 1$ and $\gamma = \text{Id}$ otherwise. For all $n \geqslant k$, the points $\gamma^{-1} s_1 \cdots s_n(0)$ and $\gamma^{-1} s'_1 \cdots s'_n(0)$ are contained in $D(s_k)$ and $D(s'_k)$ respectively. These two sets are disjoint, thus $\lim_{n \to +\infty} s_1 \cdots s_n(0) \neq \lim_{n \to +\infty} s'_1 \cdots s'_n(0)$.

This shows that f is injective.
Case 2: g_1 or g_2 is parabolic. Suppose that g_1 is parabolic. In this case, the sequences $(g_1^n(0))_{n \geqslant 1}$ and $(g_1^{-n}(0))_{n \geqslant 1}$ converge to the same point, thus f is not injective.

Let us show that $f(\Sigma_c^+) = L_c(S(g_1, g_2))$. Let $s = (s_i)_{i \geqslant 1}$ be in $(\Sigma^+ - \Sigma_c^+)$ and let k denote the smallest integer for which s_k is parabolic and $s_i = s_k$ for all $i \geqslant k$. Define $\gamma = s_1 \cdots s_{k-1}$ if $k > 1$ and $\gamma = \text{Id}$ otherwise. The point $\gamma^{-1} f(s)$ is parabolic since it is fixed by s_k, thus $f(s)$ is parabolic. As a result, the set $f(\Sigma^+ - \Sigma_c^+)$ is a subset of $L_p(S(g_1, g_2))$.

Let y be in $L_p(S(g_1, g_2))$. By Property 1.9(ii), there exists γ in $S(g_1, g_2)$ such that $\gamma(y)$ is fixed by a parabolic generator g. Let $s = (s_i)_{i \geqslant 1}$ be a sequence in Σ^+ such that $f(s) = y$. Since γ can be written as a finite reduced word, there exists $s' = (s'_i)_{i \geqslant 1}$ in Σ^+, $k \geqslant 0$ and $k' \geqslant 0$ such that $\gamma(y) = f(s')$, and for all $i \geqslant 1$, $s'_{k'+i} = s_{k+i}$. The point $\gamma(y)$ is in $D(s'_1)(\infty)$. On the other hand,

$\gamma(y)$ is the point of tangency between $D(g)$ and $D(g^{-1})$, hence $s_1' \in \{g, g^{-1}\}$. If we replace $\gamma(y)$ with $s_1^{-1}(y)$, following the same argument we obtain $s_1' = s_2'$.

Continuing this process iteratively, we find that the sequence s' must be constant. Hence for some $k \geqslant 0$, the sequence $(s_{k+i})_{i \geqslant 1}$ is constant. This implies that s is not contained in Σ_c^+. Hence $f^{-1}(L_p(S(g_1, g_2))) = (\Sigma - \Sigma_c^+)$.

In conclusion, $f(\Sigma - \Sigma_c^+) = L_p(S(g_1, g_2))$ and hence $f(\Sigma_c^+) = L_c(S(g_1, g_2))$.

Finally we must show that the function f restricted to Σ_c^+ is injective. Let $s = (s_i)_{i \geqslant 1}$ and $s' = (s_i')_{i \geqslant 1}$ be elements of Σ_c^+. Suppose that s and s' are distinct and denote by k the smallest integer $i \geqslant 1$ such that $s_i \neq s_i'$. Define $\gamma = s_1 \cdots s_{k-1}$ if $k \neq 1$ and $\gamma = \text{Id}$ otherwise.

If $s_k^{-1} \neq s_k'$, or if one of these two letters are hyperbolic, then the set $D(s_k)(\infty) \cap D(s_k')(\infty)$ is empty, thus $\gamma^{-1} f(s) \neq \gamma^{-1} f(s')$.

If $s_k^{-1} = s_k'$ and s_k is parabolic, consider the smallest $i > k$ such that $s_i \neq s_k$ and define $g = \gamma s_k \cdots s_{i-1}$. Then

$$g^{-1}(f(s)) = \lim_{n \to +\infty} s_i \cdots s_{i+n}(0),$$

$$g^{-1}(f(s')) = \lim_{n \to +\infty} s_{i-1}^{-1} \cdots s_k^{-1} s_k^{-1} s_{k+1}' \cdots s_{k+n}'(0).$$

Since $s_{k+1}' \neq s_k$, the point $g^{-1}(f(s'))$ is in $D(s_{i-1}^{-1})(\infty)$. Furthermore, $g^{-1}(f(s))$ is in $D(s_i)(\infty)$, and $D(s_i)(\infty) \cap D(s_{i-1}^{-1})(\infty) = \emptyset$, since s_i is not contained in $\{s_{i-1}, s_{i-1}^{-1}\}$. Therefore $g^{-1}(f(s')) \neq g^{-1}(f(s))$ and hence $f(s) \neq f(s')$. □

It follows that the fixed points of parabolic isometries in $L(S(g_1, g_2))$ are encoded (non-uniquely) by the sequences $(s_i)_{i \geqslant 1}$ in Σ^+ which are constant for large i, and whose repeated term is a parabolic generator.

Let us now analyze the encoding of all the fixed points of the isometries in $S(g_1, g_2)$. Consider the *shift function* $T : \Sigma^+ \to \Sigma^+$ defined by

$$T((s_i)_{i \geqslant 1}) = (s_{i+1})_{i \geqslant 1}.$$

A sequence s is *periodic* if there exists $k \geqslant 1$ such that $T^k s = s$. In this case, one writes

$$s = (\overline{s_1, \ldots, s_k}).$$

More generally, if there exists $n \geqslant 1$ such that $T^n s$ is periodic, then the sequence s is said to be *almost periodic*.

Note that if s is a sequence in $(\Sigma^+ - \Sigma_c^+)$, there exists $k \geqslant 0$ such that $T^k(s) = (\overline{s_{k+1}})$ with s_{k+1} parabolic. Hence, such a sequence is almost periodic and $f(s)$ is the fixed point of $\gamma s_{k+1} \gamma^{-1}$, where $\gamma = s_1 \cdots s_k$. The following property generalizes this connection between almost periodic sequences and fixed points of isometries of $S(g_1, g_2)$.

Property 2.3. *A point y in $L(S(g_1, g_2))$ is fixed by a non-trivial isometry in $S(g_1, g_2)$ if and only if there exists an almost periodic sequence s in Σ^+ such that $f(s) = y$.*

Proof. Let y be in $L(S(g_1, g_2))$. Suppose that there exists a non-trivial γ in $S(g_1, g_2)$ such that $y = \lim_{n \to +\infty} \gamma^n(0)$. Write γ as a reduced word $\gamma = s_1 \cdots s_n$. If $s_1 \neq s_n^{-1}$, then the periodic sequence $s = (\overline{s_1, \ldots, s_n})$ is contained in Σ^+ and $y = f(s)$.

Otherwise, consider the largest $1 \leqslant k < n$ such that $s_k = s_{n-k+1}^{-1}$, and define $g = s_1 \cdots s_k$. The point $g^{-1}(y)$ is fixed by the reduced word $s_{k+1} \cdots s_{n-k}$. Since $s_{k+1} \neq s_{n-k}^{-1}$, one has $g^{-1}(y) = f(\overline{s_{k+1}, \ldots, s_{n-k}})$. Consider the almost periodic sequence s' defined by $s_i' = s_i$ for all $1 \leqslant i \leqslant k$ and $T^k(s') = (\overline{s_{k+1}, \ldots, s_{n-k}})$. Since $s_{n-k} \neq s_{k+1}^{-1}$, this sequence is in Σ^+ and $f(s') = y$.

Conversely, consider an almost periodic sequence s in Σ^+. Let $k \geqslant 0$ be such that $T^k(s)$ is the periodic sequence $s' = (\overline{s_{k+1} \cdots s_n})$. Then

$$f(s') = \lim_{p \to +\infty} (s_{k+1} \cdots s_n)^p(0),$$

hence $f(s')$ is fixed by $\gamma = s_{k+1} \cdots s_n$. If $k = 0$, then $f(s') = f(s)$; otherwise $f(s) = g(f(s'))$ with $g = s_1 \cdots s_k$. Thus $f(s)$ is fixed by $g\gamma g^{-1}$. □

Since Γ is geometrically finite, we have $L(S(g_1, g_2)) = L_p(S(g_1, g_2)) \cup L_c(S(g_1, g_2))$. By definition of conical points, if x is a point in the set $L(S(g_1, g_2)) - L_p(S(g_1, g_2))$, then there exists a sequence $(\gamma_n)_{n \geqslant 1}$ in $S(g_1, g_2)$ such that $(\gamma_n(0))_{n \geqslant 1}$ converges to x, remaining at a bounded distance from the geodesic ray $[0, x)$.

How to construct such a sequence $(\gamma_n)_{n \geqslant 1}$? The answer is found in the coding.

Since x is not parabolic, there exists a unique sequence s in Σ_c^+ satisfying $f(s) = x$. Consider a new sequence $s' = (s_i')_{i \geqslant 1}$ constructed from s by grouping together consecutive terms corresponding to the same parabolic generator, defined by:

- $s_1' = s_1$ if s_1 is hyperbolic and $n = 1$,
- $s_1' = s_1^n$ if s_1 is parabolic, with $n \geqslant 1$ satisfying $s_1 = s_2 = \cdots = s_n$ and $s_{n+1} \neq s_1$. Such an n exists by the definition of Σ_c^+.

Repeat this procedure, beginning with the sequence $(s_{n+i})_{i \geqslant 1}$, to find s_2'. Step by step, this procedure produces a sequence $(s_i')_{i \geqslant 1}$ satisfying the following properties:

(i) $s_i' = a_i^{n_i}$ with $a_i \in \mathcal{A}$. If a_i is hyperbolic, then $n_i = 1$ and $a_{i+1} \neq a_i^{-1}$; if a_i is parabolic, then $n_i \in \mathbb{N}^*$ and $a_{i+1} \neq a_i^{\pm 1}$.

(ii) For all $i \geqslant 1$, the arcs $D(a_i)(\infty)$ and $D(a_{i+1})(\infty)$ are disjoint.

(iii) $\lim_{n \to +\infty} s_1' \cdots s_n'(0) = x$.

Note that, if g_1 and g_2 are hyperbolic, then $s = s'$.

Property 2.4. *Let x in $L_c(S(g_1, g_2))$. There exists $\varepsilon > 0$ such that the sequence $(s_1' \cdots s_n'(0))_{n \geqslant 1}$ is contained in an ε-neighborhood of the geodesic ray $[0, x)$.*

Proof. Write $\gamma_n = s'_1 \cdots s'_n$. Fix a point $y \neq x$ in $\mathcal{D}_0(S(g_1, g_2))(\infty)$.

The point $\gamma_n^{-1}(x)$ is in $D(s'_{n+1})(\infty)$. Since y does not belong to the interior of the arcs $D(a)(\infty)$ for all a in $\mathcal{A}$, it follows from Property 1.4(i) that the point $\gamma_n^{-1}(y)$ is in $D(s'_n)(\infty)$. The construction of the sequence $(s'_i)_{i \geqslant 1}$ requires the arcs $D(s'_i)(\infty)$ and $D(s'_{i+1})(\infty)$ to be disjoint. Thus the Euclidean distance between $\gamma_n^{-1}(x)$ and $\gamma_n^{-1}(y)$ is bounded below by a positive constant which does not depend on n. This property implies that there exists a compact subset of $\mathbb{D}$ whose image under γ_n intersects the geodesic (yx). Take z in the geodesic (yx). Since $\lim_{n \to +\infty} \gamma_n(0) = x$, the sequence $(\gamma_n(0))_{n \geqslant 1}$ converges to x and remains within a bounded distance from the geodesic ray $[z, x)$. It follows that there exists $\varepsilon > 0$ such that the sequence $(\gamma_n(0))_{n \geqslant 1}$ is in the ε-neighborhood of the geodesic ray $[0, x)$. $\qquad\square$

3 The modular group and two subgroups

Let us now return to the Poincaré half-plane. In this section, we study the action on $\mathbb{H}$ of the *modular group* $\mathrm{PSL}(2, \mathbb{Z})$ composed of Möbius transformations h of the form

$$h(z) = \frac{az + b}{cz + d} \quad \text{with } a, b, c, d \in \mathbb{Z} \text{ and } ad - bc = 1,$$

and of two of its subgroups.

3.1 The modular group

By definition, the modular group is Fuchsian. We are going to describe one of its Dirichlet domains.

Exercise 3.1. Prove that only the trivial isometry in $\mathrm{PSL}(2, \mathbb{Z})$ fixes the point $2i$.

Define the isometries $T_1(z) = z + 1$ and $s(z) = -1/z$. These two isometries will allow us to construct the Dirichlet domain of the modular group.

Property 3.2.

$$\mathcal{D}_{2i}(\mathrm{PSL}(2, \mathbb{Z})) = \{z \in \mathbb{H} \mid |z| \geqslant 1 \text{ and } -1/2 \leqslant \mathrm{Re}\, z \leqslant 1/2\}.$$

Proof. Set $E = \{z \in \mathbb{H} \mid |z| \geqslant 1 \text{ and } -1/2 \leqslant \mathrm{Re}\, z \leqslant 1/2\}$. By definition of the Dirichlet domain (see Sect. I.2.3), $\mathcal{D}_{2i}(\mathrm{PSL}(2, \mathbb{Z}))$ is contained in the set $\mathbb{H}_{2i}(T_1) \cap \mathbb{H}_{2i}(T_1^{-1}) \cap \mathbb{H}_{2i}(s)$. On the other hand,

$$\mathbb{H}_{2i}(T_1) = \{z \in \mathbb{H} \mid \mathrm{Re}\, z \leqslant 1/2\},$$
$$\mathbb{H}_{2i}(T_1^{-1}) = \{z \in \mathbb{H} \mid \mathrm{Re}\, z \geqslant -1/2\} \quad \text{and}$$
$$\mathbb{H}_{2i}(s) = \{z \in \mathbb{H} \mid |z| \geqslant 1\},$$

hence $\mathcal{D}_{2i}(\mathrm{PSL}(2, \mathbb{Z})$ is contained in E (Fig. II.8).

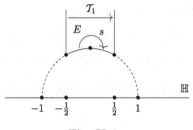

Fig. II.8.

Let z be in $\mathring{E}$. Suppose that there exists $\gamma(z) = (az + b)/(cz + d)$ in $\mathrm{PSL}(2, \mathbb{Z}) - \{\mathrm{Id}\}$ such that $\gamma(z)$ is in E. Then $c \neq 0$ necessarily since $|\mathrm{Re}(z + b)| > 1/2$ for all $b \in \mathbb{Z}^*$. One has $\mathrm{Im}(\gamma z) = \mathrm{Im}\, z/|cz + d|^2$. Also, since z is contained in $\mathring{E}$,

$$|cz + d|^2 > (|c| - |d|)^2 + |c||d|.$$

Therefore, $c \neq 0$ implies $\mathrm{Im}\, z > \mathrm{Im}(\gamma(z))$.

If $\gamma(z)$ is contained in $\mathring{E}$, the same reasoning using γ^{-1} allows us to conclude that $\mathrm{Im}(\gamma(z)) > \mathrm{Im}\, z$, a contradiction.

In conclusion, for all γ in $\mathrm{PSL}(2, \mathbb{Z}) - \{\mathrm{Id}\}$, one has $\gamma \mathring{E} \cap \mathring{E} = \varnothing$. This implies that $\mathring{E}$ is contained in $\mathcal{D}_{2i}(\mathrm{PSL}(2, \mathbb{Z}))$. Thus $\mathcal{D}_{2i}(\mathrm{PSL}(2, \mathbb{Z})) = E$. $\square$

This proposition immediately produces the following result.

Corollary 3.3. *The group* $\mathrm{PSL}(2, \mathbb{Z})$ *is a non-uniform lattice.*

Exercise 3.4. Verify that the *modular surface* $\mathrm{PSL}(2, \mathbb{Z}) \backslash \mathbb{H}$ has the form of Fig. II.9.
(Hint: use Proposition I.2.17.)

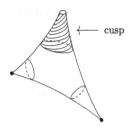

Fig. II.9.

In Sect. 4, we will use another fundamental domain constructed from $\mathcal{D}_{2i}(\mathrm{PSL}(2, \mathbb{Z}))$, as defined in the following exercise.

Exercise 3.5. Prove that the set

$$\Delta = \mathcal{D}_{2i}(\mathrm{PSL}(2,\mathbb{Z})) \cap \{z \in \mathbb{H} \mid \mathrm{Re}\, z \geqslant 0\}$$
$$\cap\, T_1(\mathcal{D}_{2i}(\mathrm{PSL}(2,\mathbb{Z})) \cap \{z \in \mathbb{H} \mid \mathrm{Re}\, z \leqslant 0\}),$$

is a fundamental domain of the modular group (Fig. II.10).

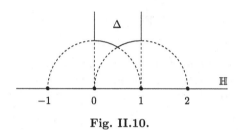

Fig. II.10.

The group $\mathrm{PSL}(2,\mathbb{Z})$, unlike Schottky groups, contains elliptic elements, as $s(z) = -1/z$ and $r(z) = (z-1)/z$.

Proposition 3.6. *An elliptic element of* $\mathrm{PSL}(2,\mathbb{Z})$ *is conjugate in* $\mathrm{PSL}(2,\mathbb{Z})$ *to some power of either s or r.*

Exercise 3.7. Prove Proposition 3.6.
(Hint: use the fact that the trace of such element is -1, 0, or 1.)

Which isometries in the modular group are parabolic? As a consequence of Corollary I.4.10, if p is such an isometry, the orbit of its fixed point intersects $\mathcal{D}_{2i}(\mathrm{PSL}(2,\mathbb{Z}))(\infty)$. This set is reduced to the point ∞, which is parabolic since it is fixed by T_1. The stabilizer of this point in $\mathrm{PSL}(2,\mathbb{Z})$ is generated by T_1, hence p is conjugate in $\mathrm{PSL}(2,\mathbb{Z})$ to a power of T_1.

From the preceding discussion, it follows that all parabolic points are of the form $\gamma(\infty)$, with γ in $\mathrm{PSL}(2,\mathbb{Z})$. If γ does not fix the point ∞, the transformation γ can be written as $\gamma(z) = (az+b)/(cz+d)$ with $c \neq 0$, thus $\gamma(\infty)$ is the rational number a/c.

Conversely, let p/q be in $\mathbb{Q}$, with $\gcd(p,q) = 1$. Consider p', q' in $\mathbb{Z}$ such that $pq' - qp' = 1$. Define $g(z) = (pz+p')/(qz+q')$. This isometry belongs to $\mathrm{PSL}(2,\mathbb{Z})$ and $g(\infty) = p/q$ thus p/q is fixed by gT_1g^{-1}.

We have proved the following facts.

Property 3.8.

(i) *The parabolic isometries of the modular group are conjugate in* $\mathrm{PSL}(2,\mathbb{Z})$ *to powers of T_1.*

(ii) $L_p(\mathrm{PSL}(2,\mathbb{Z})) = \mathbb{Q} \cup \{\infty\}$.

Therefore, the rational numbers correspond to the parabolic points of $\text{PSL}(2,\mathbb{Z})$, distinct from ∞. Furthermore, since this group is a lattice, its limit set is $\mathbb{H}(\infty)$ and is the disjoint union of the set of parabolic points and conical points. It follows that the irrational numbers correspond to conical points.

This geometric characterization of a rational is the key to the last section of Chap. VII. It allows us to relate the theory of Diophantine approximation to the behavior of geodesics on the modular surface.

3.2 Congruence modulo 2 subgroup and the commutator subgroup

One interesting aspect of these two subgroups is that they share the same fundamental domain. However, this domain is Dirichlet only in the first case.

The congruence modulo 2 subgroup. Let P be the group homomorphism of $\text{PSL}(2,\mathbb{Z})$ into $\text{SL}(2,\mathbb{Z}/2\mathbb{Z})$ sending any Möbius transformation $h(z) = (az+b)/(cz+d)$ with integer coefficients to the matrix

$$P(h) = \begin{pmatrix} a^{\bullet} & b^{\bullet} \\ c^{\bullet} & d^{\bullet} \end{pmatrix},$$

where $n^{\bullet}$ denotes the class of n in $\mathbb{Z}/2\mathbb{Z}$. The group $\Gamma(2) = P^{-1}\left(\begin{smallmatrix} 1^{\bullet} & 0^{\bullet} \\ 0^{\bullet} & 1^{\bullet} \end{smallmatrix}\right)$ is called the congruence modulo 2 subgroup [41, Chap. V.5]. This is a normal subgroup of $\text{PSL}(2,\mathbb{Z})$ of index 6.

Let r be the Möbius transformation which sends z to $r(z) = (z-1)/z$. Then

$$(*) \qquad \Gamma(2)\backslash\text{PSL}(2,\mathbb{Z}) = \{\overline{\text{Id}}, \overline{r}, \overline{r^2}, \overline{T_1^{-1}}, \overline{T_1^{-1}r}, \overline{T_1^{-1}r^2}\}.$$

Consider the set Δ' (Fig. II.11) defined by

$$\Delta' = \Delta \cup r\Delta \cup r^2\Delta \cup T_1^{-1}(\Delta) \cup T_1^{-1}r(\Delta) \cup T_1^{-1}r^2(\Delta).$$

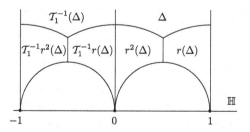

Fig. II.11.

Exercise 3.9. Prove that Δ' is a fundamental domain of $\Gamma(2)$.
(Hint: use Exercise 3.5 and the relation (*).)

The isometry $\mathcal{T}_{-1}(z) = z/(z+1)$ will useful in the following discussion.

Property 3.10. *The domain Δ' is the Dirichlet domain of $\Gamma(2)$ at the point i.*

Proof. None of the elements of $\Gamma(2) - \{\mathrm{Id}\}$ fixes i. Furthermore, the isometries $\mathcal{T}_1^{\pm 2}, \mathcal{T}_{-1}^{\pm 2}$ belong to $\Gamma(2)$, so one has

$$\mathbb{H}_i(\mathcal{T}_1^2) = \{z \in \mathbb{H} \mid \mathrm{Re}\, z \leqslant 1\},$$
$$\mathbb{H}_i(\mathcal{T}_1^{-2}) = \{z \in \mathbb{H} \mid \mathrm{Re}\, z \geqslant -1\},$$
$$\mathbb{H}_i(\mathcal{T}_{-1}^2) = \{z \in \mathbb{H} \mid |z + 1/2| \geqslant 1/2\}, \quad \text{and}$$
$$\mathbb{H}_i(\mathcal{T}_{-1}^{-2}) = \{z \in \mathbb{H} \mid |z - 1/2| \geqslant 1/2\}.$$

Therefore, $\Delta' = \mathbb{H}_i(\mathcal{T}_1^2) \cap \mathbb{H}_i(\mathcal{T}_1^{-2}) \cap \mathbb{H}_i(\mathcal{T}_{-1}^2) \cap \mathbb{H}_i(\mathcal{T}_{-1}^{-2})$. It follows that $\mathcal{D}_i(\Gamma(2))$ is contained in Δ'. However, since $\mathcal{D}_i(\Gamma(2))$ and Δ' are two fundamental domains of $\Gamma(2)$, one has $\Delta' = \mathcal{D}_i(\Gamma(2))$. □

Corollary 3.11. *The group $\Gamma(2)$ is a non-uniform lattice.*

Exercise 3.12. Verify that the surface $\Gamma(2)\backslash\mathbb{H}$ is of the form given by Fig. II.12.
(Hint: use Proposition I.2.17.)

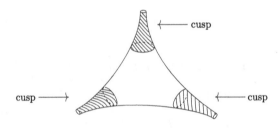

Fig. II.12.

The group $\Gamma(2)$ is a generalized Schottky group (see Sect. 1.1 for the definition). Recall that the Möbius transformation $\psi(z) = i\frac{z-i}{z+i}$ (see Sect. 1.5) sends $\mathbb{H}$ into the Poincaré disk. We have $0 = \psi(i)$. Set:

$$\mathcal{A} = \{\psi\mathcal{T}_1^2\psi^{-1}, \psi\mathcal{T}_1^{-2}\psi^{-1}, \psi\mathcal{T}_{-1}^2\psi^{-1}, \psi\mathcal{T}_{-1}^{-2}\psi^{-1}\}.$$

For all a in $\mathcal{A}$, the half-planes $D(a)$ bounded by the perpendicular bisectors of the hyperbolic segments $[0, a(0)]_h$ are either tangent or disjoint. From Proposition 1.6, the group generated by $\mathcal{T}_1^2$ and $\mathcal{T}_{-1}^2$ is therefore free (and discrete) and it has Δ' as its fundamental domain. Since this group is a subset of $\Gamma(2)$ and has the same fundamental domain, they are equal.

Thus we have the following property:

Property 3.13. *The group* $\Gamma(2)$ *is generated by* T_1^2 *and* T_{-1}^2, *and is free relative to these generators.*

We now consider the parabolic isometries in $\Gamma(2)$.

Property 3.14. *The parabolic isometries of* $\Gamma(2)$ *are conjugate to powers of* T_1^2, T_{-1}^2 *or* $T_{-1}^{-2}T_1^2$ *in* $\Gamma(2)$.

Proof. Notice first that the point ∞ is fixed by T_1^2, the point 0 is fixed by T_{-1}^2, the point -1 is fixed by $T_{-1}^{-2}T_1^2$, and the point 1 is fixed by $T_{-1}^2T_1^{-2}$. These four points are parabolic. Furthermore, -1 and 1 are in the same orbit since $T_1^2(-1) = 1$, and the sets $\Gamma(2)(0), \Gamma(2)(\infty), \Gamma(2)(1)$ are three disjoint orbits.

Let γ be a parabolic isometry in $\Gamma(2)$. From Corollary I.4.10, its fixed point is contained in one of the three orbits described above. Also since each of $T_1^2, T_{-1}^{-2}T_1^2, T_{-1}^2$ generates the stabilizer in $\Gamma(2)$ of its fixed point, γ is conjugate to a power of one of these three isometries. $\qquad\square$

Note that, unlike Schottky groups, the parabolic isometries of a generalized Schottky group admitting a collection of non-elliptic generators $\{g_1, \ldots, g_p\}$ satisfying the condition $(D(g_i) \cup D(g_i^{-1})) \cap (D(g_j) \cup D(g_j^{-1})) = \varnothing$, for all $i \neq j$ in $\{1, \ldots, p\}$ are not always conjugate to powers of the parabolic generators g_i.

Exercise 3.15. Prove that the set $L_p(\Gamma(2))$ is equal to $\mathbb{Q} \cup \{\infty\}$.
(Hint: use Property 3.8 and the fact that $\Gamma(2)$ is normal in $\mathrm{PSL}(2, \mathbb{Z})$, (see also [41, Chap. V, Example F]).)

The commutator subgroup. We now introduce another subgroup of the modular group defined this time by a given collection of generators. Let α_1 and α_2 be two Möbius transformations defined as

$$\alpha_1(z) = \frac{z+1}{z+2}, \qquad \alpha_2(z) = \frac{z-1}{-z+2},$$

and consider the half-planes

$$B(\alpha_1) = \{z \in \mathbb{H} \mid |z - 1/2| \leqslant 1/2\}, \qquad B(\alpha_1^{-1}) = \{z \in \mathbb{H} \mid \mathrm{Re}\, z \leqslant -1\},$$
$$B(\alpha_2) = \{z \in \mathbb{H} \mid |z + 1/2| \leqslant 1/2\}, \qquad B(\alpha_2^{-1}) = \{z \in \mathbb{H} \mid \mathrm{Re}\, z \geqslant 1\}.$$

For all $i = 1, 2$ and $\varepsilon = \pm 1$, one has

$$\alpha_i^\varepsilon(\mathbb{H} - \overset{\circ}{B}(\alpha_i^{-\varepsilon})) = B(\alpha_i^\varepsilon).$$

Note that we have (Fig. II.13) the following relation:

$$\Delta' = \bigcap_{\substack{\varepsilon = \pm 1 \\ i = 1,2}} \mathbb{H} - \overset{\circ}{B}(\alpha_i^{-\varepsilon}).$$

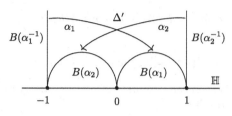

Fig. II.13.

Exercise 3.16.

(i) Prove that if the perpendicular bisector of a hyperbolic segment $[z_1, z_2]_h$ is a vertical half-line, then $\operatorname{Im} z_1 = \operatorname{Im} z_2$.
(Hint: use Exercise I.1.8.)

(ii) Conclude that there is no point z in $\mathbb{H}$ such that the half-planes $B(\alpha_1^{-1})$ and $B(\alpha_2^{-1})$ are bounded by the perpendicular bisectors of the segments $[z, \alpha_1^{-1}(z)]_h$ and $[z, \alpha_2^{-1}(z)]_h$ respectively.

Let Γ be the group generated by α_1 and α_2. The group Γ being a subgroup of $\mathrm{PSL}(2, \mathbb{Z})$, it is Fuchsian. Observe that Γ is not a generalized Schottky group with respect to α_1 and α_2, since there is no $z \in \mathbb{D}$ such that $B(\alpha_i^\varepsilon) = D_z(\alpha_i^\varepsilon)$. Even so, the arguments presented in the first part of the proof of Proposition 1.6 being purely dynamic, they still apply.

Property 3.17. *The group Γ generated by α_1 and α_2 is free relative to these generators. Moreover Γ is the commutator subgroup of $\mathrm{PSL}(2, \mathbb{Z})$ (i.e., it is generated by the elements $[g, h] = ghg^{-1}h^{-1}$, where g and h are in $\mathrm{PSL}(2, \mathbb{Z})$).*

Exercise 3.18. Prove Property 3.17.
(Hint: For the first part of Property 3.17, see proof of Proposition 1.6. For the second part, use the identities $[s, T_1^{-1}] = \alpha_1$ and $[s, T_1] = \alpha_2$.)

Exercise 3.19. Prove that Γ is a normal subgroup of index 6 in $\mathrm{PSL}(2, \mathbb{Z})$. Moreover, give the structure of the two finite groups $\mathrm{PSL}(2, \mathbb{Z})/\Gamma(2)$ and $\mathrm{PSL}(2, \mathbb{Z})/\Gamma$.

Let us prove that the set Δ' is a fundamental domain of Γ. Notice that, since Γ is not a generalized Schottky group, the second part of Proposition 1.6 does not apply directly.

Property 3.20. *The set Δ' is a fundamental domain of Γ.*

Proof. Define $\mathcal{A} = \{\alpha_1^{\pm 1}, \alpha_2^{\pm 1}\}$. Let us show that, for all z in $\mathbb{H}$, there exists γ in Γ such that $\gamma(z)$ belongs to Δ'.

If z is not in Δ', there exists a_1 in $\mathcal{A}$ such that z belongs to $B(a_1)$. Define $z_1 = a_1^{-1}(z)$. If z_1 is contained in Δ', one may define $z_n = z_1$ for all $n \geqslant 1$. Otherwise there exists a_2 in $\mathcal{A} - \{a_1^{-1}\}$ such that z_1 belongs to $B(a_2)$ and one may define $z_2 = a_2^{-1}(z_1)$. Following this process, we construct the sequence $(z_n)_{n \geqslant 1}$.

If $(z_n)_{n \geqslant 1}$ is constant after some N, then $a_N^{-1} \cdots a_1^{-1}(z)$ is contained in Δ'. Otherwise, define $\gamma_n = a_1 \cdots a_n$. By construction, $\gamma_n^{-1}(z)$ is contained in $B(a_{n+1})$. Consider a subsequence $(\gamma_{n_p})_{p \geqslant 1}$ such that $a_{n_p+1} = a$. The point z belongs to $\gamma_{n_p}(B(a))$ for all p. For all $p \geqslant 2$, the half-plane $\gamma_{n_p}(B(a))$ is contained in $\gamma_{n_{p-1}}(B(a))$ (see the argument in Property 1.4(ii)). The point z lies in each of these half-planes, hence there is a compact K in $\mathbb{H}$ such that the image by $a_1 \cdots a_{n_p}$ of the geodesic bounding $B(a)$ meets K, for all $p \geqslant 1$. This geodesic is a subset of Δ' and $\Delta' = \Delta \cup r\Delta \cup r^2 \Delta \cup T_1^{-1}\Delta \cup T_1^{-1}r\Delta \cup T_1^{-1}r^2\Delta$. Therefore, there exist infinitely many elements γ in $\mathrm{PSL}(2, \mathbb{Z})$ such that $\gamma\Delta$ intersects K. This contradicts the fact that Δ is a locally finite fundamental domain of $\mathrm{PSL}(2, \mathbb{Z})$ (since it is a Dirichlet domain). Thus the sequence $(z_n)_{n \geqslant 1}$ is necessarily constant and there exists γ in Γ such that $\gamma(z) \in \Delta'$.

Furthermore, for any non-trivial γ in Γ, the open set $\gamma(\overset{\circ}{\Delta}')$ is a subset of some open half-plane $\overset{\circ}{B}(a)$ with $a \in \mathcal{A}$ (see the argument from Property 1.4(i)), thus its intersection with $\overset{\circ}{\Delta}'$ is empty.   □

Exercise 3.21. Verify that the surface $\Gamma \backslash \mathbb{H}$ is of the form given by Fig. II.14. (Hint: use the fact that since Δ' is locally finite, and that the function from Δ' modulo Γ to in $\Gamma \backslash \mathbb{H}$, sending $\Gamma z \cap \Delta'$ to Γz, is a homeomorphism (see Proposition I.2.17).)

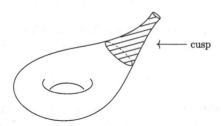

Fig. II.14.

Since the domain Δ' is not a Dirichlet domain, we cannot use the results of Chap. I to conclude that Γ is a non-uniform lattice. This property is in fact true, but requires proof. Our chosen proof is not especially direct, but has the advantage of illustrating some interesting properties of the group and of using Criterion I.4.17: Γ is a non-uniform lattice if and only if $L(\Gamma) = \mathbb{H}(\infty)$ and $L(\Gamma) = L_p(\Gamma) \cup L_h(\Gamma)$, with $L_p(\Gamma) \neq \varnothing$.

Let us consider the non-trivial translation

$$[\alpha_2^{-1}, \alpha_1^{-1}] = \alpha_2^{-1}\alpha_1^{-1}\alpha_2\alpha_1.$$

Exercise 3.22. Prove that $[\alpha_2^{-1}, \alpha_1^{-1}]$ generates the stabilizer of the point ∞ in Γ.

Property 3.23. *The parabolic isometries of Γ are conjugate to powers of $[\alpha_2^{-1}, \alpha_1^{-1}]$ in Γ.*

Proof. Fix z in $\overset{\circ}{\Delta}'$. Consider a parabolic isometry γ in Γ and write it in the form of a reduced word $s_1 \cdots s_n$, where $\mathcal{A} = \{\alpha_1^{\pm 1}, \alpha_2^{\pm 1}\}$. One may assume that $s_1 \neq s_n^{-1}$. In this case, $\lim_{k \to +\infty} \gamma^k(i)$ is contained in $B(s_1)$ and $\lim_{k \to +\infty} \gamma^{-k}(i)$ is contained in $B(s_n^{-1})$. These two limits are equal to the unique fixed point x of γ, thus x is an element of $\{-1, 0, 1, \infty\}$. If $x = \infty$, then γ belongs to the group generated by $[\alpha_2^{-1}, \alpha_1^{-1}]$. Otherwise, since $1 = \alpha_1(\infty)$, $-1 = \alpha_2(\infty)$, and $0 = \alpha_2^{-1}\alpha_1(\infty)$, the element γ is conjugate to a power of $[\alpha_2^{-1}, \alpha_1^{-1}]$. $\square$

Exercise 3.24. Let H be a non-elementary Fuchsian group and N be a normal subgroup. Prove that $L(H) = L(N)$.
(Hint: use the minimality of the limit set.)

Property 3.25. *The group Γ is a non-uniform lattice and $L_p(\Gamma) = \mathbb{Q} \cup \{\infty\}$.*

Proof. The group Γ is normal in $\mathrm{PSL}(2, \mathbb{Z})$, thus one has $L(\Gamma) = \mathbb{H}(\infty)$. Let us examine $L_p(\Gamma)$. We know that the point ∞ is contained in this set. Furthermore, for all γ in $\mathrm{PSL}(2, \mathbb{Z})$, the Möbius transformation $\gamma [\alpha_2^{-1}, \alpha_1^{-1}] \gamma^{-1}$ is a parabolic isometry in Γ which fixes $\gamma(\infty)$. Thus $L_p(\Gamma) = \mathbb{Q} \cup \{\infty\}$ (Property 3.8).

Consider now an irrational number x. This point is horocyclic with respect to $\mathrm{PSL}(2, \mathbb{Z})$, hence there exists a sequence $(\gamma_n)_{n \geqslant 1}$ in the modular group satisfying

$$\lim_{n \to +\infty} B_x(z, \gamma_n(z)) = +\infty.$$

Since Γ has finite index in $\mathrm{PSL}(2, \mathbb{Z})$, passing to a subsequence one has $\gamma_n = g_n \gamma$, where g_n is in Γ and $\lim_{n \to +\infty} B_x(z, g_n(z)) = +\infty$. Thus x is a horocyclic point with respect to Γ.

In conclusion,

$$L(\Gamma) = \mathbb{H}(\infty), \quad L_p(\Gamma) = \mathbb{Q} \cup \{\infty\} \quad \text{and} \quad L(\Gamma) = L_p(\Gamma) \cup L_h(\Gamma). \quad \square$$

4 Expansions of continued fractions

We have shown in the preceding section that $\mathbb{Q} \cup \{\infty\}$ is the set of parabolic points associated to the modular group. In this section, we continue to weave the relationships between number theory and hyperbolic geometry, by creating a geometric context for representations of irrational numbers x as continued fractions.

We begin by recalling the algorithmic definition of this representation. We define $x_0 = x$ and $n_0 = E(x_0)$, where $E(x)$ designates the integer part of x. For all $i \geqslant 1$, define x_i and n_i by the following recurrence relation:

$$x_i = 1/(x_{i-1} - n_{i-1}) \quad \text{and} \quad n_i = E(x_i).$$

Let $[n_0; n_1, \ldots, n_k]$ denote the rational number defined by

$$[n_0; n_1, \ldots, n_k] = n_0 + \cfrac{1}{n_1 + \cfrac{1}{n_2 + \cfrac{\ddots}{\quad + \cfrac{1}{n_{k-1} + \cfrac{1}{n_k}}}}}$$

The sequence of rational numbers $([n_0; n_1, \ldots, n_k])_{k \geqslant 1}$ converges to x (see [42]). The *continued fraction expansion* of x is by definition the expression of x as $[n_0; n_1, \ldots]$.

4.1 Geometric interpretation of continued fraction expansions

Consider the hyperbolic triangle T having infinite vertices at the points $\infty, 1, 0$. This triangle is related to the fundamental domain Δ of $\mathrm{PSL}(2, \mathbb{Z})$ introduced in Exercise 3.5 and defined to be

$$\Delta = \{z \in \mathbb{H} \mid 0 \leqslant \operatorname{Re} z \leqslant 1, \ |z| \geqslant 1 \text{ and } |z - 1| \geqslant 1\}.$$

More precisely, if r denotes the transformation defined by $r(z) = (z - 1)/z$, one has (see Fig. II.15)

$$T = \Delta \cup r\Delta \cup r^2 \Delta.$$

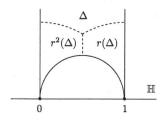

Fig. II.15.

It follows that

$$\bigcup_{\gamma \in \mathrm{PSL}(2,\mathbb{Z})} \gamma T = \mathbb{H} \quad \text{and if} \quad \gamma \mathring{T} \cap \mathring{T} \neq \varnothing, \quad \text{then } \gamma \in \{\mathrm{Id}, r, r^2\}.$$

This tiling of $\mathbb{H}$ by images of T is called the *Farey tiling*. Let L denote the unoriented geodesic whose endpoints are 0 and ∞. The images of L under PSL$(2, \mathbb{Z})$ are called *Farey lines*.

Recall that

$$T_1(z) = z + 1 \quad \text{and} \quad T_{-1}(z) = z/(z+1).$$

The endpoints of $T_1(L)$ are the points 1, ∞. Those of $T_{-1}(L)$ are 0, 1. Thus the edges of T and of γT for γ in PSL$(2, \mathbb{Z})$, are Farey lines (Fig. II.16).

Property 4.1. *Let L^+ be the geodesic L oriented from 0 to ∞. For any oriented Farey lines (xy), there exists an unique γ in* PSL$(2, \mathbb{Z})$*, such that $(xy) = \gamma(L^+)$.*

Proof. By definition, given an oriented Farey lines (xy), there exists γ in PSL$(2, \mathbb{Z})$ such that $\gamma(L)$ is the unoriented Farey line whose endpoints are x and y. If $\gamma(0) = x$ and $\gamma(\infty) = y$, then $\gamma(L^+) = (xy)$. Otherwise, $\gamma s(L^+) = (xy)$, where $s(z) = -1/z$.

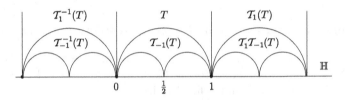

Fig. II.16.

It follows that there exists g in PSL$(2, \mathbb{Z})$ satisfying $g(L^+) = (xy)$. This element is unique since PSL$(2, \mathbb{Z})$ does not contain any non-trivial isometry fixing the points 0 and ∞. □

Definition 4.2. Let $n > 1$. A collection of n oriented Farey lines $L_1^+, \ldots, L_n^+$ are called *consecutive*, if there exists γ in PSL$(2, \mathbb{Z})$ and ε in $\{\pm 1\}$ such that for all $1 \leqslant i \leqslant n$

$$L_i^+ = \gamma T_\varepsilon^i L^+.$$

Set $L_i^+ = (x_i y_i)$. Suppose that $L_1^+, \ldots, L_n^+$ are consecutive. If $\varepsilon = 1$, then $y_i = \gamma(\infty)$ for all $1 \leqslant i \leqslant n$; likewise if $\varepsilon = -1$, then $x_i = \gamma(0)$ for all $1 \leqslant i \leqslant n$.

Figures II.17, II.18, II.19 show several cases in which three oriented Farey lines are consecutive.

If x is an element of $\mathbb{H}(\infty)$, the irrationality of x is characterized by the number of Farey lines which intersect the geodesic ray $[i, x)$.

Proposition 4.3. *Let x be in $\mathbb{H}(\infty)$. The ray $[i, x)$ intersects finitely many Farey lines if and only if x is in $\mathbb{Q} \cup \{\infty\}$.*

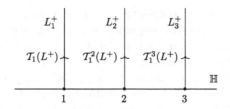

Fig. II.17.

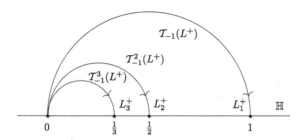

Fig. II.18.

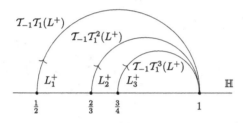

Fig. II.19.

Proof. Suppose that x belongs to $\mathbb{Q} \cup \{\infty\}$. In this case, after Property 3.8(ii), there exists γ in $\mathrm{PSL}(2,\mathbb{Z})$ such that $\gamma(x) = \infty$. Since the ray $\gamma^{-1}([i,x))$ is a vertical half-line passing through $\gamma^{-1}(i)$, there exists n in $\mathbb{Z}$ and z in $[i,x)$ such that the ray $T_1^n \gamma^{-1}([z,x))$ is in T. The domain Δ is locally finite (since it is constructed from a finite number of subsets of a Dirichlet domain (Property 3.2)), thus $T_1^n \gamma^{-1}([i,z]_h)$ intersects only a finite number of images of T under $\mathrm{PSL}(2,\mathbb{Z})$.

It follows that $T_1^n \gamma^{-1}([i,x))$—and thus $[i,x)$—only intersects finitely many Farey lines.

Suppose now that $[i,x)$ only intersects a finite number of Farey lines. Then there exists z in $[i,x)$ and γ in $\mathrm{PSL}(2,\mathbb{Z})$ such that $[z,x)$ is contained in $\gamma(T)$. In other words, $[\gamma^{-1}(z), \gamma^{-1}(x))$ is a geodesic ray contained in T. Hence, $\gamma^{-1}(x)$ is in $\{0,1,\infty\}$ and thus x is in $\mathbb{Q} \cup \{\infty\}$. $\qquad\square$

From now on, we focus on positive irrational numbers. Let x be such a real number and $r : [0, +\infty) \to [i,x)$ be the arclength parametrization of $[i,x)$.

According to the previous proposition, $[i, x)$ crosses infinitely many Farey lines. Let $(L_n)_{n \geqslant 1}$ denote the sequence of Farey lines crossed by $(r(t))_{t>0}$ in order by increasing t. For each n, let $L_n^+ = (x_n y_n)$ be the orientation on L_n defined by: at the point of intersection $r(t_n)$ between $[i, x)$ and L_n, the oriented angle from $[r(t_n), x)$ to $[r(t_n), y_n)$ belongs to $(0, \pi)$. If L_n^+ is not a vertical half-line, then $L_n^+ = (x_n y_n)$ with $x_n < y_n$, if L_n^+ is vertical, then $y_n = \infty$ (Fig. II.20). Denote by γ_n the unique element of $\mathrm{PSL}(2, \mathbb{Z})$ such that (Property 4.1)

$$\gamma_n(L^+) = L_n^+.$$

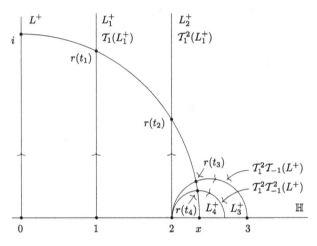

Fig. II.20.

Consider the geodesic ray $\gamma_n^{-1}([i, x))$. This ray crosses L^+ at the point $\gamma_n^{-1}(r(t_n))$. By construction, at their point of intersection $\gamma_n^{-1}(r(t_n))$, the angle oriented from $([\gamma_n^{-1}(r(t_n)), \gamma_n^{-1}(x))$ to $[\gamma_n^{-1}(r(t_n)), \infty)$ belongs to $(0, \pi)$. Thus $\gamma_n^{-1}([i, x))$ meets the Farey lines $T_{-1}(L^+) = (01)$ or $T_1(L^+) = (1\infty)$ and hence, $\gamma_n^{-1}(L_{n+1}^+)$ is equal to $T_{-1}(L^+)$ or $T_1(L^+)$. It follows that $\gamma_{n+1} = \gamma_n T_{\varepsilon_n}$, where $\varepsilon_n = \pm 1$.

Set

$$n_0 = \begin{cases} \max\{n \geqslant 1 \mid \forall k \in [1, n], \ \varepsilon_k = 1\} & \text{if } \varepsilon_1 = 1, \\ 0 & \text{if } \varepsilon_1 = -1, \end{cases}$$

$$n_p = \max\{n > n_{p-1} \mid \forall k \in (n_{p-1}, n], \ \varepsilon_k = (-1)^p\}, \quad \text{if } p \geqslant 1.$$

Note that n_k is a positive integer for all $k \geqslant 1$.

Exercise 4.4. Prove that, for all $k \geqslant 1$, the positive integer n_k represents the largest integer $p \geqslant 1$ such that $L_{n_{p-1}+1}^+, \dots, L_{n_{p-1}+p}^+$ are oriented consecutive Farey lines.

By construction of the integers n_k, we have for all $k \geqslant 1$:

$$\gamma_{n_k} = T_1^{n_0} \cdots T_{(-1)^k}^{n_k}.$$

Moreover, notice that, for $k \geqslant 1$, the intervals of $\mathbb{R}$ with extremities $\gamma_{n_k}(0)$ and $\gamma_{n_k}(\infty)$ are nested. Set $g_k = \gamma_{n_k}$. The next exercise relates the rational number $[n_0; n_1, \ldots, n_k]$, which was defined at the beginning of this section, to the point $g_k(0)$.

Exercise 4.5. Let $k \geqslant 2$. Prove that if k is even, then $g_k(0) = [n_0; n_1, \ldots, n_k]$ and $g_k(\infty) = [n_0; n_1 \cdots n_{k-1}]$. Also if k is odd, then $g_k(0) = [n_0; n_1, \ldots, n_{k-1}]$ and $g_k(\infty) = [n_0; n_1 \cdots n_k]$.
(Hint: use the relations $T_1^n(z) = z + n$ and $T_{-1}^n(z) = 1/(n + 1/z)$.)

The following proposition shows that $[n_0; n_1, \ldots]$ is the continued fraction expansion of x.

Proposition 4.6. *The sequence of rational numbers $([n_0; n_1, \ldots, n_k])_{k \geqslant 1}$ converges to x. In addition, if there exists a sequence $(n_k')_{k \geqslant 1}$ satisfying $n_0' \in \mathbb{N}$, $n_k' \in \mathbb{N}^*$ for all $k \geqslant 1$ and $\lim_{k \to +\infty} [n_0'; n_1', \ldots, n_k'] = x$, then $n_k = n_k'$ for all $k \geqslant 0$.*

Proof. The geodesic $g_k(L)$ intersects the geodesic ray $[i, x)$. By construction, the point x belongs to the interval of $\mathbb{R}$ having endpoints $g_k(0)$, $g_k(\infty)$, and these intervals are nested. For all $k \geqslant 1$, the rational numbers $g_k(0)$ and $g_k(\infty)$ belong to the interval $[n_0, n_0 + 1]$ (Exercise 4.5), thus $0 < |g_k(0) - g_k(\infty)| \leqslant 1$. Let us show that $\lim_{k \to +\infty} |g_k(0) - g_k(\infty)| = 0$. Suppose that there exists $d > 0$ and a subsequence $(g_{k_p})_{p \geqslant 1}$ such that $|g_{k_p}(0) - g_{k_p}(\infty)| > d$. In this case, the geodesic $g_{k_p}(L)$ intersects the Euclidean segment I in $\mathbb{H}$ whose endpoints are $n_0 + id/2$ and $n_0 + 1 + id/2$. Since L is an edge of T, and since T is the finite union of images of Δ, there exist infinitely many isometries γ in $\mathrm{PSL}(2, \mathbb{Z})$ such that $\gamma\Delta$ intersects the compact set I. This contradicts the fact that Δ is locally finite (since it is a finite union of subsets of a Dirichlet domain). One concludes from this property that the sequence $([n_0; n_1, \ldots, n_k])_{k \geqslant 1}$ converges to x.

Let us now show uniqueness. Suppose that $([n_0'; n_1', \ldots, n_k'])_{k \geqslant 1}$ converges to x. Then

$$\lim_{p \to +\infty} T_1^{n_0'} \cdots T_1^{n_{2p}'}(0) = \lim_{p \to +\infty} T_1^{n_0} \cdots T_1^{n_{2p}}(0).$$

The rational number $T_1^{n_0'} \cdots T_1^{n_{2p}'}(0)$ is in $(n_0', n_0' + 1)$ and $T_1^{n_0} \cdots T_1^{n_{2p}}(0)$ is in $(n_0, n_0 + 1)$. Thus $n_0' = n_0$. It follows that,

$$\lim_{p \to +\infty} T_{-1}^{n_1'} \cdots T_1^{n_{2p}'}(0) = \lim_{p \to +\infty} T_{-1}^{n_1} \cdots T_1^{n_{2p}}(0).$$

Applying the same reasoning to the sequences

$$\left(\frac{1}{[0; n_1', \ldots, n_{2p}']}\right)_{p \geqslant 1} = ([n_1'; n_2', \ldots, n_{2p}'])_{p \geqslant 1} \quad \text{and}$$

$$\left(\frac{1}{[0; n_1, \ldots, n_{2p}]_{p \geqslant 1}}\right) = ([n_1; n_2, \ldots, n_{2p}])_{p \geqslant 1},$$

one obtains $n_1 = n_1'$. Iteratively, one has $n_k = n_k'$ for all $k \geqslant 0$.   □

In summary, to find the continued fraction expansion of a positive irrational number x in terms of hyperbolic geometry, it suffices to identify x with a point in $\mathbb{H}(\infty)$, to associate to the ray $[i, x)$ the infinite sequence of oriented Farey lines $(L_i^+ = (x_i y_i))_{i \geqslant 1}$ in the order in which this ray crosses them per the procedure described above, and to count the maximal number of consecutive oriented Farey lines. Then

- $n_0 = E(x)$, and n_1 is defined by: $x_{n_0+1} = \cdots = x_{n_1} = n_0$, $x_{n_1+1} \neq n_0$;
- for all $k \geqslant 1$:
  - if k is even, then n_k is defined by $y_{n_{k-1}} = y_{n_{k-1}+1} = \cdots = y_{n_k}$ and $y_{n_k+1} \neq y_{n_k}$,
  - if k is odd, then n_k is defined by $x_{n_{k-1}} = x_{n_{k-1}+1} = \cdots = x_{n_k}$ and $x_{n_k+1} \neq x_{n_k}$.

Examples 4.7. Verify that in the situation of Fig. II.21, $n_0 = 2$, $n_1 = 2$, and $n_2 \geqslant 2$.

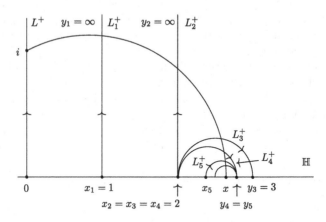

Fig. II.21.

Let S^+ denote the set of integer sequences defined by

$$S^+ = \{(n_i)_{i \geqslant 0} \mid n_0 \in \mathbb{N}, n_i \in \mathbb{N}^* \text{ for } i \geqslant 1\}.$$

Consider the function $F : \mathbb{R}^+ - \mathbb{Q}^+ \to S^+$, which sends x to the sequence $(n_i)_{i \geqslant 0}$ corresponding to the continued fraction expansion of x.

Property 4.8. *The map $F : \mathbb{R}^+ - \mathbb{Q}^+ \to S^+$ is bijective.*

Proof. According to Proposition 4.6, it is enough to show that F is surjective. For any sequence $(n_k)_{k \geqslant 0}$ in S^+, set $g_k = T_1^{n_0} \circ \cdots \circ T_{(-1)^k}^{n_k}$. Reusing the argument from the proof of Proposition 4.6, one obtains $\lim_{k \to +\infty} |g_k(0) - g_k(\infty)| = 0$. Moreover the intervals with extremities $g_k(0)$ and $g_k(\infty)$ are nested. More precisely, if k is even, then $g_{k+1}(\infty) = g_k(\infty)$, and $g_{k+1}(0)$ is in the segment having endpoints $g_k(\infty), g_k(0)$. If k is odd, then $g_{k+1}(0) = g_k(0)$ and $g_{k+1}(\infty)$ is in the segment having endpoints $g_k(\infty), g_k(0)$. Thus the sequences $(g_k(0))_{k \geqslant 1}$ and $(g_k(\infty))_{k \geqslant 1}$ converge to the same positive real number x. Exercise 4.5 implies that $x = \lim_{k \to +\infty} [n_0; n_1, n_2, \ldots, n_k]$. The real x is irrational since $[i, x)$ intersects each $g_k(L)$ (Property 4.3). $\square$

We have restricted ourselves to positive irrational number. However, if y is a negative irrational, one may identify it with a sequence of integers $(m_i)_{i \geqslant 0}$, such that $m_0 = E(y) \in \mathbb{Z}$ and $(m_i)_{i \geqslant 1} = F(y - m_0)$.

In this way, one obtains a bijection between the set of irrational numbers with the set of sequences of integers whose terms are positive—with the possible exception of the first.

Let us come back to the geometry. Since the modular group is a non-uniform lattice, its limit set is the disjoint union of parabolic points and conical points. We know that conical points correspond to irrational numbers. In the same spirit as Property 2.4 for Schottky groups, let us prove that the continued fraction expansion of an irrational number x is related to a sequence $(\gamma_k(i))_{k \geqslant 0}$ with γ_k in $\mathrm{PSL}(2, \mathbb{Z})$, remaining at a bounded distance from the geodesic ray $[i, x)$. It suffices to prove this relation when x is positive.

Property 4.9. *Let x be a positive irrational number. Set $F(x) = (n_i)_{i \geqslant 0}$ and $\gamma_k = T_1^{n_0} \cdots T_{-1}^{n_{2k+1}}$ for all $k \geqslant 0$. The sequence $(\gamma_k(i))_{k \geqslant 0}$ remains at a uniformly bounded distance from the geodesic ray $[i, x)$.*

Proof. Let $s(z) = -1/z$. Note that the point i belongs to the geodesic $(s(x) x)$. Since $-1/x < 0$ and $\operatorname{Re} \gamma_k(i) > 0$, proving Property 4.9 amounts to proving that the sequence $(\gamma_k(i))_{k \geqslant 0}$ remains at a uniformly bounded distance from the geodesic $(s(x)x)$, and hence that the sequence of couples of points $((\gamma_k^{-1}(s(x)), \gamma_k^{-1}(x)))_{k \geqslant 0}$ is contained in a compact subset of $\mathbb{H}(\infty) \times \mathbb{H}(\infty)$ minus its diagonal (Proposition I.3.14).

The continued fraction expansion of $\gamma_k^{-1}(x)$ is $[n_{2k+2}; n_{2k+3}, \ldots]$. Since n_{2k+2} is non-zero, the real $\gamma_k^{-1}(x)$ is greater than 1. Furthermore, since $s T_1^{-1} s = T_{-1}$ and $s T_{-1}^{-1} s = T_1$, one has

$$s \gamma_k^{-1}(s(x)) = T_1^{n_{2k+1}} T_{-1}^{n_{2k}} \cdots T_{-1}^{n_0}(x).$$

Hence, the continued fraction expansion of the real $s \gamma_k^{-1} s(x)$ is

$$\begin{cases} [n_{2k+1}; n_{2k}, \ldots, n_0, n_0, n_1, n_2, \ldots] & \text{if } n_0 \neq 0, \\ [n_{2k+1}; n_{2k}, \ldots, n_1, n_1, n_2, \ldots] & \text{otherwise.} \end{cases}$$

Since $n_{2k+1} \neq 0$, in both cases one obtains that the real $\gamma_k^{-1}(s(x))$ belongs to $(-1,0)$. It follows that the geodesics $((\gamma_k^{-1}(x')\gamma_k^{-1}(x)))_{k \geqslant 0}$ remain at a bounded distance from i. □

4.2 Application to the hyperbolic isometries of the modular group

We focus here on positive irrational numbers such that the sequence $F(x) = (n_i)_{i \geqslant 0}$ is almost periodic (i.e., for some $k \geqslant 0$, the sequence $(n_{k+i})_{i \geqslant 0}$ is periodic). As with the coding of the conical points of a Schottky group, we are going to show that almost periodic sequences encode the fixed points of hyperbolic isometries of the modular group.

Property 4.10.

(i) *A positive irrational number x is fixed by a hyperbolic isometry in $PSL(2, \mathbb{Z})$ if and only if the sequence $F(x)$ is almost periodic.*

(ii) *An isometry in $PSL(2, \mathbb{Z})$ is hyperbolic if and only if it is conjugate in $PSL(2, \mathbb{Z})$ to an isometry of the form $T_1^{m_1} T_{-1}^{m_2} \cdots T_{-1}^{m_k}$ with $m_i > 0$ and k even.*

Proof.

(i) Let x be a positive irrational number. Suppose that the sequence $F(x) = (n_i)_{i \geqslant 0}$ is periodic, in which case n_0 is non-zero. Let T denote the period of this sequence, and define $k = T - 1$ if T is even, and $k = 2T - 1$ otherwise. Then $x = \lim_{p \to +\infty}(T_1^{n_0} \cdots T_1^{n_k})^p(0)$. This shows that x is fixed by $T_1^{n_0} \cdots T_{-1}^{n_k}$, which is hyperbolic since x is irrational.

If $F(x)$ is almost periodic, after q initial terms for some $q \geqslant 1$, it suffices to apply the preceding reasoning to the point $(T_1^{n_0} \cdots T_{-1}^{n_q})^{-1}(x)$ if q is odd, and to the point $(T_1^{n_0} \cdots T_{-1}^{n_{q-1}})^{-1}(x)$ if q is even.

Consider now a hyperbolic isometry γ in $PSL(2, \mathbb{Z})$. Let $F(\gamma^+) = (n_i)_{i \geqslant 0}$ and set $g_k = T_1^{n_0} \cdots T_{(-1)^k}^{n_k}$. Recall that the geodesic ray $[i, x)$ meets all oriented Farey lines $L_k^+ = (g_k(0)g_k(\infty))$. According to Exercise 4.5, the sequences $(g_k(0))_{k \geqslant 0}$ and $(g_k(\infty))_{k \geqslant 0}$ converge to γ^+. Thus for large enough k, the Euclidean segment having endpoints $g_k(0), g_k(\infty)$ does not contain γ^-, and hence there exists $k' > k$ such that $\gamma L_k^+ = L_{k'}^+$. Applying Property 4.1, one obtains that $\gamma g_k = g_{k'}$. It follows that $\gamma = g_k T_{(-1)^{(k+1)}}^{n_{k+1}} \cdots T_{(-1)^{k'}}^{n_k'} g_k^{-1}$. If k and k' are both odd or even, then the sequence $F(g_k^{-1}(\gamma^+))$ is periodic. Otherwise, $F(T_{(-1)^{(k+1)}}^{-n_{k+1}} g_k^{-1}(\gamma^+))$ is periodic. In both cases, the sequence $F(\gamma^+)$ is almost periodic.

(ii) Let γ be a hyperbolic isometry in $PSL(2, \mathbb{Z})$. After conjugating γ by a translation, one may suppose that $\gamma^+ > 0$. According to the end of the proof of part (i), γ is conjugate to $T_1^{n_{k+1}} \cdots T_{(-1)^{k'}}^{n_k'}$ if k and k' are both odd or even, and to $\mathcal{T}_1^{n_{k+2}} \cdots \mathcal{T}_{(-1)^{k'}}^{n_k'}$, otherwise. Hence it is conjugate to an isometry of the form $\mathcal{T}_1^{m_1} \cdots \mathcal{T}_{-1}^{m_p}$, with $m_i > 0$ and p even. Conversely, an isometry of the form $\mathcal{T}_1^{m_1} \mathcal{T}_{-1}^{m_2} \cdots \mathcal{T}_{-1}^{m_k}$ with $m_i > 0$ and k even, is hyperbolic since it fixes $\lim_{p \to +\infty} (\mathcal{T}_1^{m_1} \cdots \mathcal{T}_{-1}^{m_k})^p(0)$ and $\lim_{p \to -\infty} (\mathcal{T}_1^{m_1} \cdots \mathcal{T}_{-1}^{m_k})^p(0)$, which are distinct real numbers.   □

This property allows us to establish a relationship between the fixed points of hyperbolic isometries of the modular group and the *quadratic real numbers*, which are solutions to equations like

$$ax^2 + bx + c = 0 \quad \text{with } a \in \mathbb{N}^* \text{ and } b, c \in \mathbb{Z}.$$

Proposition 4.11. *Let x be an irrational number. Then the following are equivalent:*

(i) *x is fixed by a hyperbolic isometry in* $\mathrm{PSL}(2, \mathbb{Z})$;
(ii) *x is quadratic.*

We give a proof of this well-known result (see for example [42]) using the transformations $\mathcal{T}_1$ and $\mathcal{T}_{-1}$.

Proof. The implication from (i) to (ii) requires only two facts. The first one is that a fixed point x of a Möbius transformation $\gamma(z) = (az + b)/(cz + d)$ satisfies the Diophantine equation

$$Ax^2 + Bx - C = 0,$$

with $A = c, B = d - a$ and $C = -b$. The second one is that the integer c is non-zero since γ is hyperbolic.

Let us now prove that (ii) implies (i). Let α and β be two distinct roots of an equation of the form

$$Ax^2 + Bx - C = 0,$$

with $A \neq 0$ and B, C in $\mathbb{Z}$.

We want to show that these two real numbers are fixed by some hyperbolic isometry of the modular group. After replacing them by $g(\alpha)$ and $g(\beta)$, where g is in $\mathrm{PSL}(2, \mathbb{Z})$, one may assume that $\alpha > 0$ and $\beta < 0$. Hence $A > 0$ and $C > 0$. Set $F(\alpha) = (n_i)_{i \geq 0}$. For all even integer $k > 0$, define the real numbers

$$x_k = (\mathcal{T}_1^{n_0} \cdots \mathcal{T}_{-1}^{n_{k-1}})^{-1}(\alpha) \quad \text{and} \quad y_k = (\mathcal{T}_1^{n_0} \cdots \mathcal{T}_{-1}^{n_{k-1}})^{-1}(\beta),$$

and set $x_0 = \alpha, y_0 = \beta$.

We have $x_k = \lim_{p \to +\infty} [n_k; n_{k+1}, \ldots, n_{k+p}]$, hence x_k is positive. Furthermore, an induction argument shows that y_k is negative and that the two real numbers x_k and y_k are solutions of an equation of the form

$$A_k x^2 + B_k x - C_k = 0,$$

where $A_k, B_k, C_k \in \mathbb{Z}$, $A_k > 0$, $C_k > 0$ and $B_k^2 + 4A_kC_k = B^2 + 4AC$. Thus the coefficients A_k, B_k, C_k belong to a finite set. It follows that there exist two even integers $k_2 > k_1 \geqslant 0$ such that $A_{k_1} = A_{k_2}, B_{k_1} = B_{k_2}, C_{k_1} = C_{k_2}$. This implies that $x_{k_1} = x_{k_2}$, and hence

$$T_1^{n_{k_1}} \cdots T_{-1}^{n_{k_2}-1}(x_{k_2}) = x_{k_2}.$$

We obtain that the real number x_{k_2} is the fixed point of the hyperbolic isometry $g' = T_1^{n_{k_1}} \cdots T_{-1}^{n_{k_2}-1}$. Since $\alpha = T_1^{n_0} \cdots T_{-1}^{n_{k_2}-1}(x_{k_2})$, this real is fixed by a conjugate of g'. The same reasoning applied to y_{k_2} implies that β is similarly fixed by the same hyperbolic isometry. $\quad\square$

We conclude this section by focusing on the displacements of isometries in $\mathrm{PSL}(2, \mathbb{Z})$. Recall that the displacement $\ell(\gamma)$ of an isometry γ is defined (see I.2.2) by

$$\ell(\gamma) = \inf_{z \in \mathbb{H}} d(z, \gamma(z)).$$

Exercise 4.12. Let $\gamma(z) = (az + b)/(cz + d)$ be a hyperbolic isometry in G. Denote λ the eigenvalue of the matrix $\left(\begin{smallmatrix} a & b \\ c & d \end{smallmatrix}\right)$ whose absolute value is > 1. Prove the equality

$$\ell(\gamma) = 2 \ln |\lambda|.$$

The following property relates the fixed point of a hyperbolic isometry in $\mathrm{PSL}(2, \mathbb{Z})$ to its displacement. Let $\sigma : (\mathbb{R} - \mathbb{Q}) \cap (1, +\infty) \to (\mathbb{R} - \mathbb{Q}) \cap (1, +\infty)$ defined by

$$\sigma(x) = \frac{1}{x - n_0},$$

where n_0 is the first term of the sequence $F(x)$. Notice that the sequence $F(\sigma(x))$ is the shifted sequence $(n_{i+1})_{i \geqslant 0}$.

Property 4.13. *If γ is an isometry of the form $\gamma = T_1^{m_1} T_{-1}^{m_2} \cdots T_{-1}^{m_k}$, with m_i in $\mathbb{N}^*$ and k even, then*

$$\ell(\gamma) = 2 \ln(\gamma^+ \times \sigma(\gamma^+) \times \cdots \times \sigma^{k-1}(\gamma^+)).$$

Proof. Set $\gamma(z) = (az + b)/(cz + d)$, $M = \left(\begin{smallmatrix} a & b \\ c & d \end{smallmatrix}\right)$ and $\lambda = e^{\ell(\gamma)/2}$. We have

$$M \begin{pmatrix} \gamma^+ \\ 1 \end{pmatrix} = \lambda \begin{pmatrix} \gamma^+ \\ 1 \end{pmatrix}.$$

Consider the matrices $D_n = \left(\begin{smallmatrix} 0 & 1 \\ 1 & n \end{smallmatrix}\right)$ and $R = \left(\begin{smallmatrix} 0 & 1 \\ 1 & 0 \end{smallmatrix}\right)$. These matrices satisfy the following relations:

$$R^2 = \mathrm{Id}, \quad D_n R = \begin{pmatrix} 1 & 0 \\ n & 1 \end{pmatrix} \quad \text{and} \quad RD_n = \begin{pmatrix} 1 & n \\ 0 & 1 \end{pmatrix}.$$

Using these relations, one obtains

$$\lambda\begin{pmatrix}1\\ \gamma^+\end{pmatrix} = D_{m_1}\cdots D_{m_k}\begin{pmatrix}1\\ \gamma^+\end{pmatrix}.$$

If $x \neq 0$, then $D_m\left(\frac{1}{x}\right) = x\left(\frac{1}{m+1/x}\right)$. Therefore, $D_{m_k}\left(\frac{1}{\gamma^+}\right) = \gamma^+\left(\frac{1}{m_k+1/\gamma^+}\right)$. However, $m_k + 1/\gamma^+ = \sigma^{k-1}(\gamma^+)$, hence

$$\lambda\begin{pmatrix}1\\ \gamma^+\end{pmatrix} = \gamma^+ D_{m_1}\cdots D_{m_{k-1}}\begin{pmatrix}1\\ \sigma^{k-1}(\gamma^+)\end{pmatrix}.$$

Repeating this process, one obtains

$$\lambda\begin{pmatrix}1\\ \gamma^+\end{pmatrix} = \gamma^+ \sigma^{k-1}(\gamma^+)\cdots\sigma^2(\gamma^+) D_{m_1}\begin{pmatrix}1\\ \sigma(\gamma^+)\end{pmatrix}.$$

Furthermore, $D_{m_1}\left(\frac{1}{\sigma(\gamma^+)}\right) = \sigma(\gamma^+)\left(\frac{1}{\gamma^+}\right)$, thus $\lambda = \gamma^+ \prod_{i=1}^{k-1}\sigma^i(\gamma^+)$.   □

Using this property, one obtains an interpretation—in terms of hyperbolic geometry—of the *golden ratio*

$$\mathcal{N} = \frac{1 + \sqrt{5}}{2}.$$

Corollary 4.14. *If γ is a hyperbolic isometry in $\mathrm{PSL}(2,\mathbb{Z})$, then*

$$\ell(\gamma) \geqslant 2\ln(\mathcal{T}_1\mathcal{T}_{-1})^+ = 4\ln\mathcal{N}.$$

Proof. Suppose γ is a hyperbolic isometry in $\mathrm{PSL}(2,\mathbb{Z})$. According to Property 4.10(ii), we can suppose that this isometry is of the form $\mathcal{T}_1^{m_1}\cdots\mathcal{T}_{-1}^{m_k}$, with $m_i > 0$. Let us prove that $\ell(\mathcal{T}_1^{m_1}\cdots\mathcal{T}_{-1}^{m_k}) > 4\ln\mathcal{N}$, if some $m_i \neq 1$. Let x be the attractive fixed point of this isometry, the sequence $F(x)$ is the periodic sequence $m_1,\ldots,m_k,m_1,\ldots,m_k,m_1,\ldots$. For $1 \leqslant i \leqslant k$, notice that the real $\sigma^i(x)$ is of the form $m_{i+1} + 1/(m_{i+2} + x_i)$, where $0 < x_i < 1$. Therefore, if one of the m_i is equal to 2, there exist j,l with $0 \leqslant j,l \leqslant k-1$ and $j \neq l$ such that $\sigma^j(x) = 2 + 1/y$, where $y > 1$, and $\sigma^l(x) = m_{l+1} + 1/(2 + x_l)$, with $0 < x_l < 1$. From these remarks and Property 4.13, one obtains

$$\ell(\gamma) > 2\ln\left(2 \times (1 + 1/3)\right) = 2\ln 8/3 > 4\ln\mathcal{N}.$$

If one of the m_i is $\geqslant 3$, then $\ell(\gamma) \geqslant 2\ln 3 > 4\ln\mathcal{N}$.

Suppose now that all of the m_i are 1, then $\mathcal{T}_1^{m_1}\cdots\mathcal{T}_{-1}^{m_k}$ is a power of $\mathcal{T}_1\mathcal{T}_{-1}$. The attractor x of $\mathcal{T}_1\mathcal{T}_{-1}$ satisfies $x = 1 + 1/x$, hence $x = \mathcal{N}$ and, after Property 4.13, one has $\ell(\mathcal{T}_1\mathcal{T}_{-1}) = 2\ln\mathcal{N}^2$.

In conclusion, $\ell(\mathcal{T}_1^{m_1}\cdots\mathcal{T}_{-1}^{m_k}) > 4\ln\mathcal{N}$, except if $k = 2$ and $m_1 = m_2 = 1$.   □

Corollary 4.14 has a geometric interpretation on the modular surface $S = \mathrm{PSL}(2,\mathbb{Z})\backslash\mathbb{H}$. We will see in Chap. III that the set of all $\ell(\gamma)$, where γ is a hyperbolic isometry of $\mathrm{PSL}(2,\mathbb{Z})$, is the set of the lengths of all compact geodesics in the surface S (see Sect. III.3 and Appendix B for the notion of a geodesic on S). In this context, Corollary 4.14 says that the real $4\ln\mathcal{N}$ corresponds to the length of the shortest compact geodesic on the modular surface.

5 Comments

The construction of Schottky groups and the coding of their limit sets that have been introduced in this chapter can be generalized to pinched Hadamard manifolds X (see the Comments in Chap. I) whose group of positive isometries contains at least two non-elliptic elements g_1, g_2 having no common fixed points. Under this condition, for large enough n_0, there exist two disjoint compact subsets K_1 and K_2 of $X(\infty)$ satisfying the following relation for all $i = 1, 2$ and $|n| \geqslant n_0$:

$$g_i^n(X(\infty) - K_i) \subset K_i.$$

An application of the *Ping-Pong Lemma* [36] shows that the group generated by $g_1^{n_0}, g_2^{n_0}$ is a free Kleinian group. Such a group is geometrically finite [21]. Without any other hypotheses on X, these groups—which are again called *Schottky groups*—and their variants are in general the only accessible non-elementary Kleinian groups.

On the other hand, if one restricts to the case in which X is a symmetric space of rank 1, one can construct Kleinian groups (in general, lattices) using number theory.

In the particular case of the Poincaré half-plane, the arithmetic groups are known [41, Chap. 5]. The modular groups and $\Gamma(2)$ belong to this rich family. Let us mention (see [45]) a rather unexpected example of a lattice contained in the group of Möbius transformations having rational coefficients which is not commensurable to the modular group but for which the set of parabolic points is still $\mathbb{Q} \cup \{\infty\}$.

The geometric construction of continued fraction expansions that was introduced in this chapter is essentially taken from two articles [57, 20]. In addition to these two references, the text of C. Series in [8] also studies the limit set of Schottky groups and the modular group, but goes further toward the construction of a coding of the limit set of an arbitrary geometrically finite Fuchsian group.

III

Topological dynamics of the geodesic flow

In this chapter we focus on the geodesic flow on the quotient of $T^1\mathbb{H}$ by a Fuchsian group. Our motivation is to give relations between the behavior of this flow and the nature of the points in the limit set of the group. General notions related to topological dynamics are introduced in Appendix A.

1 Preliminaries on the geodesic flow

1.1 The geodesic flow on $T^1\mathbb{H}$

Recall from Proposition I.1.10 that $T^1\mathbb{H}$ is equipped with the G-invariant distance D which is defined by

$$D((z, \overrightarrow{v}), (z', \overrightarrow{v}')) = \int_{-\infty}^{+\infty} e^{-|t|} d(v(t), v'(t)) \, dt,$$

where $(v(t))_{t\in\mathbb{R}}$ represents the arclength parametrization of the unique oriented geodesic passing through z whose tangent line at z is in the direction of $\overrightarrow{v}$, and which satisfies

$$v(0) = z \quad \text{and} \quad \frac{dv}{dt}(0) = \overrightarrow{v}.$$

Let $v(-\infty)$, $v(+\infty)$ the points in $\mathbb{H}(\infty)$ corresponding respectively to the negative and positive endpoints of this geodesic (Fig. III.1).

Exercise 1.1. Let $((z_n, \overrightarrow{v_n}))_{n\geqslant 1}$ be a sequence in $T^1\mathbb{H}$. Prove that

$$\lim_{n\to+\infty} D((z_n, \overrightarrow{v_n}), (z, \overrightarrow{v})) = 0$$

$$\Longleftrightarrow \quad \lim_{n\to+\infty} v_n(+\infty) = v(+\infty) \quad \text{and} \quad \lim_{n\to+\infty} z_n = z.$$

F. Dal'Bo, *Geodesic and Horocyclic Trajectories*, Universitext,
DOI 10.1007/978-0-85729-073-1_3, © Springer-Verlag London Limited 2011

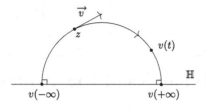

Fig. III.1.

We now define the geodesic flow on $T^1\mathbb{H}$. Let $\tilde{g}$ be the function from $\mathbb{R} \times T^1\mathbb{H}$ into $T^1\mathbb{H}$ defined by

$$\tilde{g}(t', (z, \overrightarrow{v})) = \left(v(t'), \frac{d}{dt}v(t')\right).$$

Exercise 1.2. Prove that for all real numbers t the function $\tilde{g}_t$ is a homeomorphism of $T^1\mathbb{H}$ into itself.

Note that for all t, t' in $\mathbb{R}$ one has

$$D(\tilde{g}_t(z, \overrightarrow{v}), \tilde{g}_{t'}(z, \overrightarrow{v})) = 2|t - t'|.$$

It follows from this remark and Exercise 1.2 that the function $\tilde{g}$ is continuous.

Exercise 1.3. Prove that $\tilde{g}_{t+t'} = \tilde{g}_t \circ \tilde{g}_{t'}$, for all t, t' in $\mathbb{R}$.

Thus the function $\tilde{g}$ is a well-defined flow on $T^1\mathbb{H}$. This flow is called the *geodesic flow*. Its dynamics is analogous to the dynamics of the flow on $\mathbb{R}^2$ associated to a non-zero vector field (Example A.2(i) in Appendix A).

Exercise 1.4. Prove the following properties:

(i) the non wandering set (Definition A.11 in Appendix A) $\Omega_{\tilde{g}}(T^1\mathbb{H})$ is empty;
(ii) all points in $T^1\mathbb{H}$ are divergent points.

1.2 The geodesic flow on a quotient

By analogy with the flow on the torus $\mathbb{T}^2$, viewed as the quotient of the Euclidean plane by the translations group $\mathbb{Z}^2$, induced by a linear flow on $\mathbb{R}^2$ (Examples A.2(i) and (ii) in Appendix A), the geodesic flow on $T^1\mathbb{H}$ induces a flow on the quotient of this space by a Fuchsian group.

More precisely, consider a Fuchsian group Γ and let π (respectively π^1) be the projection of $\mathbb{H}$ (respectively $T^1\mathbb{H}$) onto the quotient $S = \Gamma\backslash\mathbb{H}$ (respectively $T^1S = \Gamma\backslash T^1\mathbb{H}$) (Figs. III.2 and III.3). Each of these quotients is equipped with a distance defined respectively by

$$d_\Gamma(\pi(x), \pi(y)) = \inf_{\gamma \in \Gamma} d(x, \gamma(y)),$$

$$D_\Gamma(\pi^1(z, \overrightarrow{u}), \pi^1(z', \overrightarrow{u}')) = \inf_{\gamma \in \Gamma} D((z, \overrightarrow{u}), \gamma(z', \overrightarrow{u}')).$$

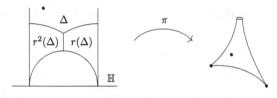

Fig. III.2. $\varGamma = \mathrm{PSL}(2,\mathbb{Z})$

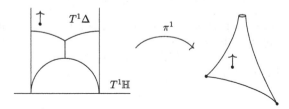

Fig. III.3. $\varGamma = \mathrm{PSL}(2,\mathbb{Z})$

Exercise 1.5. Prove that $d_\varGamma$ and $D_\varGamma$ are distance functions and that the topologies induced by these distances on S and T^1S are the same as those induced by π and π^1.

The notion of convergence of a sequence can be interpreted in S and T^1S in the following ways:

(i) A sequence $(\pi(z_n))_{n\geqslant 1}$ in S converges to $\pi(z)$ if and only if there exists a sequence $(\gamma_n)_{n\geqslant 1}$ in $\varGamma$ such that $(\gamma_n(z_n))_{n\geqslant 1}$ converges to z.
(ii) A sequence $(\pi^1((z_n, \overrightarrow{u_n})))_{n\geqslant 1}$ in T^1S converges to $\pi^1((z, \overrightarrow{u}))$ if and only if there exists a sequence $(\gamma_n)_{n\geqslant 1}$ in $\varGamma$ such that $(\gamma_n((z_n, \overrightarrow{u_n})))_{n\geqslant 1}$ converges to $(z, \overrightarrow{u})$.

If $\varGamma$ does not have any elliptic element, then the surface S is a differentiable manifold whose Riemannian structure is induced by that of $\mathbb{H}$, and T^1S is its unitary tangent bundle. This is not the case if there are elliptic elements in $\varGamma$. The group $\mathrm{PSL}(2,\mathbb{Z})$ is an example of such a group. When $\varGamma = \mathrm{PSL}(2,\mathbb{Z})$, the π-projection of a hyperbolic disk, of sufficiently small radius, centered at $j = 1/2 + i\sqrt{3}/2$ is homeomorphic to the cone obtained by taking the quotient of this disk by the cyclic group of order 3 generated by $r(z) = (z - 1)/z$. Therefore, in a neighborhood of $\pi(j)$, the modular surface does not inherit the manifold structure of $\mathbb{H}$. The same is true in a neighborhood of $\pi(i)$. In this context one cannot talk about the Riemannian structure in the classical sense. Therefore, despite its misleading notation, T^1S is not always the unitary tangent bundle of S. Whether or not $\varGamma$ has elliptic elements, we will show that the flow $\widetilde{g}$ induces a flow on the topological space T^1S.

Let $(z, \overrightarrow{v})$ in $T^1\mathbb{H}$ and $(v(t))_{t\in\mathbb{R}}$ be the arclength parametrization of the oriented geodesic $(\gamma(v(-\infty))\gamma(v(+\infty)))$, such that $v(0) = z$. For any positive isometry $\gamma \in \varGamma$, the function $\gamma \circ v : \mathbb{R} \to \mathbb{H}$ which sends t to $\gamma(v(t))$ is the arclength parametrization of the oriented geodesic $(\gamma(v(-\infty))\gamma(v(+\infty)))$ satisfying

$$\gamma(v(0)) = \gamma(z) \quad \text{and} \quad \frac{d}{dt}\gamma \circ v(0) = T_z\gamma(\overrightarrow{v}).$$

It follows that $\tilde{g}_t(\gamma(z, \overrightarrow{v})) = \gamma(\tilde{g}_t((z, \overrightarrow{v})))$, for all $t \in \mathbb{R}$. This last relation allows us to define the *geodesic flow* $g : \mathbb{R} \times T^1S \to T^1S$ (Fig. III.4) by

$$g(t, \pi^1((z, \overrightarrow{v}))) = \pi^1(\tilde{g}(t, (z, \overrightarrow{v}))).$$

The rest of this chapter is devoted to the topological dynamics of this flow. We use—especially in Sects. 3 and 4—the following convergence criterion relating the action of $g_\mathbb{R}$ on $\Gamma\backslash T^1\mathbb{H}$ to the dual action of Γ on the set of oriented geodesics $\tilde{g}_\mathbb{R}\backslash T^1\mathbb{H}$.

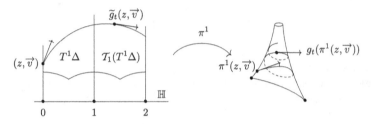

Fig. III.4. $\Gamma = \mathrm{PSL}(2, \mathbb{Z})$

Proposition 1.6. *Let* $((z_n, \overrightarrow{u_n}))_{n\geqslant 1}$ *be a sequence in* $T^1\mathbb{H}$ *and* $(z, \overrightarrow{u})$ *be an element of* $T^1\mathbb{H}$. *The following properties are equivalent:*

(i) *there exists a sequence of real numbers* $(s_n)_{n\geqslant 1}$ *such that*

$$\lim_{n\to+\infty} g_{s_n}(\pi^1((z_n, \overrightarrow{u_n}))) = \pi^1((z, \overrightarrow{u}));$$

(ii) *there exists a sequence* $(\gamma_n)_{n\geqslant 1}$ *in* Γ *such that*

$$\lim_{n\to+\infty} (\gamma_n(u_n(-\infty)), \gamma_n(u_n(+\infty))) = (u(-\infty), u(+\infty)).$$

Proof.
  (i) $\Rightarrow$ (ii). By definition of the topology on T^1S, there exists a sequence $(\gamma_n)_{n\geqslant 1}$ in Γ such that

$$\lim_{n\to+\infty} D(\gamma_n\tilde{g}_{s_n}((z_n, \overrightarrow{u_n})), (z, \overrightarrow{u})) = 0.$$

This convergence together with Exercise 1.1 implies that the ordered pair

$$(\gamma_n(u_n(-\infty)), \gamma_n(u_n(+\infty)))$$

converges to $(u(-\infty), u(+\infty))$.

(ii) $\Rightarrow$ (i). Consider the sequence of oriented geodesics

$$L_n = (\gamma_n(u_n(-\infty))\gamma_n(u_n(+\infty))).$$

This sequence converges to the geodesic $L = (u(-\infty)u(+\infty))$, hence there exists a sequence of points z'_n in L_n converging to z. Let $(z'_n, \overrightarrow{v_n})$ be the element of $T^1\mathbb{H}$ such that $\overrightarrow{v_n}$ is the unit vector at z'_n which is tangent to the ray $[z'_n, \gamma_n(u_n(+\infty)))$. There exists $s_n \in \mathbb{R}$ such that

$$(z'_n, \overrightarrow{v_n}) = \tilde{g}_{s_n}(\gamma_n((z_n, \overrightarrow{u_n}))).$$

Since $\lim_{n\to+\infty} z'_n = z$ and $\lim_{n\to+\infty} v_n(+\infty) = u(+\infty)$, by Exercise 1.1 one has $\lim_{n\to+\infty} g_{s_n}(\pi^1((z_n, \overrightarrow{u_n}))) = \pi^1((z, \overrightarrow{u}))$. $\square$

2 Topological properties of geodesic trajectories

We fix a non-elementary Fuchsian group Γ. The motivation of this chapter is to study the behavior of the trajectories of the geodesic flow g on T^1S. We use the notions introduced in Appendix A.

2.1 Characterization of the wandering and divergent points

Since Γ is not elementary group, its limit set $L(\Gamma)$ is minimal (Proposition I.3.6). The following theorem gives a characterization of the non-wandering set $\Omega_g(T^1S)$ (Appendix A) of the geodesic flow on T^1S in terms of points in $L(\Gamma)$.

Theorem 2.1. Let $(z, \overrightarrow{u})$ be in $T^1\mathbb{H}$. Then the following are equivalent:

(i) $\pi^1((z, \overrightarrow{u}))$ belongs to $\Omega_g(T^1S)$;
(ii) $u(-\infty)$ and $u(+\infty)$ belong to $L(\Gamma)$.

Before we prove this theorem, we will prove the following lemma.

Lemma 2.2. Let x, y be points in $L(\Gamma)$. There exists a sequence $(\gamma_n)_{n\geqslant 1}$ in Γ such that $\lim_{n\to+\infty} \gamma_n(i) = x$ and $\lim_{n\to+\infty} \gamma_n^{-1}(i) = y$.

Proof. Fix some x in $L(\Gamma)$. Let A denote the set of x' in $L(\Gamma)$ for which there exists a sequence $(h_n)_{n\geqslant 1}$ in Γ satisfying $\lim_{n\to+\infty} h_n(i) = x$ and $\lim_{n\to+\infty} h_n^{-1}(i) = x'$. This set is non-empty and Γ-invariant. We will show that it is also closed.

Let $(x'_p)_{p\geqslant 1}$ be a sequence in A converging to a point x' in $\mathbb{H}(\infty)$. For all p, there exists a sequence $(h_{p,k})_{k\geqslant 1}$ in Γ such that $\lim_{k\to+\infty} h_{p,k}(i) = x$ and $\lim_{k\to+\infty} h_{p,k}^{-1}(i) = x'_p$. Therefore, there exists a sequence $(h_{p,k_p})_{p\geqslant 1}$ satisfying $\lim_{p\to+\infty} h_{p,k_p}(i) = x$ and $\lim_{p\to+\infty} h_{p,k_p}^{-1}(i) = x'$. This implies that the point x' is in A and therefore that A is closed. Since A is a non-empty closed subset of $L(\Gamma)$, and $L(\Gamma)$ is minimal, we have $A = L(\Gamma)$. $\square$

Proof (of Theorem 2.1).

(ii) $\Rightarrow$ (i). Let $(\gamma_n)_{n\geqslant 1}$ be in Γ such that $\lim_{n\to+\infty}\gamma_n(i) = u(+\infty)$ and $\lim_{n\to+\infty}\gamma_n^{-1}(i) = u(-\infty)$. Define $t_n = d(z,\gamma_n^{-1}(z))$. The sequence $(t_n)_{n\geqslant 0}$ converges to $+\infty$. Consider the element $(z_n, \overrightarrow{v_n})$ in $T^1\mathbb{H}$, where $z_n = \gamma_n^{-1}(z)$ and $\overrightarrow{v_n}$ is the unit vector tangent to the segment $[z_n, z]_h$ at z_n (Fig. III.5).

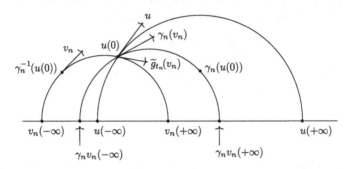

Fig. III.5.

One has $\lim_{n\to+\infty} v_n(-\infty) = u(-\infty)$ and $\lim_{n\to+\infty} v_n(+\infty) = u(+\infty)$. Furthermore $v_n(t_n) = z$ and therefore $\lim_{n\to+\infty}\widetilde{g}_{t_n}((z_n,\overrightarrow{v_n})) = (z, \overrightarrow{u})$.

Consider now $\gamma_n((z_n, \overrightarrow{v_n}))$. This element corresponds to the ordered pair composed of the point z and the unit vector tangent to the geodesic segment $[z, \gamma_n(z)]_h$ at z. Observe that $\lim_{n\to+\infty}\gamma_n((z_n, \overrightarrow{v_n})) = (z, \overrightarrow{u})$. Let V be a neighborhood of $\pi^1((z, \overrightarrow{u}))$. For large enough n, $\pi^1((z_n,\overrightarrow{v_n}))$ and $g_{t_n}(\pi^1((z_n,\overrightarrow{v_n})))$ belong to V, thus $g_{t_n}V \cap V \neq \varnothing$. This shows that $\pi^1((z, \overrightarrow{u}))$ is non-wandering.

(i) $\Rightarrow$ (ii). Let $(V_n)_{n\geqslant 1}$ be a sequence of neighborhoods of $\pi^1((z, \overrightarrow{u}))$ such that $\bigcap_{n=1}^{+\infty} V_n = \{\pi^1((z, \overrightarrow{u}))\}$. Since $\pi^1((z, \overrightarrow{u}))$ is non-wandering, there exists a sequence $t_n \to +\infty$ such that $g_{t_n}V_n \cap V_n \neq \varnothing$. From this remark, it follows that there exists a sequence $(\pi^1((z_n, \overrightarrow{u_n})))_{n\geqslant 1}$ in T^1S satisfying

$$\lim_{n\to+\infty} \pi^1((z_n, \overrightarrow{u_n})) = \pi^1((z, \overrightarrow{u})) \quad \text{and} \quad \lim_{n\to+\infty} g_{t_n}\pi^1((z_n, \overrightarrow{u_n})) = \pi^1((z, \overrightarrow{u})).$$

Replacing $(z_n, \overrightarrow{u_n})$ by an element of $\Gamma((z_n, \overrightarrow{u_n}))$, there exists a sequence $(\gamma_n)_{n\geqslant 1}$ in Γ satisfying

$$\lim_{n\to+\infty} (z_n, \overrightarrow{u_n}) = (z, \overrightarrow{u}) \quad \text{and} \quad \lim_{n\to+\infty} \gamma_n\widetilde{g}_{t_n}((z_n, \overrightarrow{u_n})) = (z, \overrightarrow{u}).$$

Since $\lim_{n\to+\infty} t_n = +\infty$, one has $\lim_{n\to+\infty} u_n(t_n) = u(+\infty)$. Furthermore $\lim_{n\to+\infty} d(u_n(t_n), \gamma_n^{-1}z) = 0$, thus $\lim_{n\to+\infty} \gamma_n^{-1}(i) = u(+\infty)$. This shows that $u(+\infty)$ belongs to $L(\Gamma)$.

Replacing $(t_n)_{n\geqslant 1}$ by $(-t_n)_{n\geqslant 1}$ in the preceding argument, it is clear that $u(-\infty)$ also belongs to $L(\Gamma)$. $\square$

Corollary 2.3. *The set $\Omega_g(T^1S)$ is equal to T^1S if and only if $L(\Gamma) = \mathbb{H}(\infty)$.*

The following proposition characterizes the fact that the set $\Omega_g(T^1S)$ is compact in terms of points in $L(\Gamma)$.

Proposition 2.4. *The set $\Omega_g(T^1S)$ is compact if and only if all points of $L(\Gamma)$ are conical.*

Before we prove Proposition 2.4, we introduce the subset $\widetilde{\Omega}_g(T^1S) \subset T^1\mathbb{H}$ defined by $\widetilde{\Omega}_g(T^1S) = (\pi^1)^{-1}\Omega_g(T^1S)$. It follows from Theorem 2.1 that

$$\widetilde{\Omega}_g(T^1S) = \{(z, \overrightarrow{u}) \in T^1\mathbb{H} \mid u(-\infty) \in L(\Gamma), \ u(+\infty) \in L(\Gamma)\}.$$

Proof. The projection to $\mathbb{H}$ of $\widetilde{\Omega}_g(T^1S)$ is the set

$$\widetilde{\Omega}(\Gamma) = \{z \in \mathbb{H} \mid z \in (xy) \text{ with } x, y \in L(\Gamma)\}$$

(this set was introduced in Sect. I.4.1). Note that $\Omega_g(T^1S)$ is compact if and only if there exists a compact $K \subset \mathbb{H}$ such that $\widetilde{\Omega}(\Gamma) = \bigcup_{\gamma \in \Gamma} \gamma K$.

If every point in $L(\Gamma)$ is conical, Corollary I.4.17 implies that the group Γ is convex-cocompact. By definition the group Γ acts on the convex-hull of $\widetilde{\Omega}(\Gamma)$ with a compact fundamental domain and hence that $\widetilde{\Omega}(\Gamma) = \bigcup_{\gamma \in \Gamma} \gamma K$, for some compact $K \subset \mathbb{H}$.

Conversely, suppose that there exists a compact subset $K \subset \mathbb{H}$ such that $\widetilde{\Omega}(\Gamma) = \bigcup_{\gamma \in \Gamma} \gamma K$. For any geodesic (xy) in $\widetilde{\Omega}(\Gamma)$, there exists a sequence of points on this geodesic of the form $(\gamma_n(k_n))_{n \geqslant 1}$ converging to x, where $\gamma_n \in \Gamma$ and $k_n \in K$. Fix z on (xy). The sequence $(\gamma_n(z))_{n \geqslant 1}$ remains within a bounded distance of the ray $[z, x)$. Thus x is conical. $\square$

In the particular case where Γ is a Schottky group $S(g_1, g_2)$ (see Chap. II), we obtain that the set $\Omega_g(T^1S)$ is compact if and only if g_1 and g_2 are hyperbolic (Figs. III.6 and III.7).

Fig. III.6. $\Gamma = S(g_1, g_2)$, g_1 and g_2 hyperbolic

Recall that a point $y \in Y$ is divergent (respectively positively or negatively divergent) for a flow ϕ on Y, if for all unbounded sequences $(t_n)_{n \geqslant 1}$ in $\mathbb{R}$ (respectively $\mathbb{R}^+$ or $\mathbb{R}^-$), the sequence $(\phi_{t_n}(y))_{n \geqslant 0}$ diverges (see Appendix A).

Fig. III.7. $\Gamma = S(g_1, g_2)$, g_1 hyperbolic and g_2 parabolic

Let us analyze the divergent points for the geodesic flow on $T^1 S$. Notice that we can restrict our attention to the positively divergent points. To prove it, we introduce the *flip map* on each unitary tangent plane $T^1_z \mathbb{H}$, which associates to $(z, \overrightarrow{u})$, the point $-(z, \overrightarrow{u}) = (z, -\overrightarrow{u})$.

Exercise 2.5. Prove that the flip map is continuous, and that for all $t \in \mathbb{R}$ and $\gamma \in G$, one has

$$-\widetilde{g}_t(-(z, \overrightarrow{u})) = \widetilde{g}_{-t}((z, \overrightarrow{u})) \quad \text{and} \quad \gamma(-(z, \overrightarrow{u})) = -\gamma((z, \overrightarrow{u})).$$

Using this exercise, we obtain

Lemma 2.6. *The point $\pi^1((z, \overrightarrow{u}))$ is positively divergent if and only if $\pi^1(-(z, \overrightarrow{u}))$ is negatively divergent.*

Suppose now that $\pi^1((z, \overrightarrow{u}))$ is a positively divergent point. There exist a positive unbounded sequence $(t_n)_{n \geqslant 0}$ and a sequence $(\gamma_n)_{n \geqslant 0}$ in Γ such that $(\widetilde{g}_{t_n} \gamma_n (z, \overrightarrow{u}))_{n \geqslant 0}$ converges to some $(z', \overrightarrow{u}')$ in $T^1 \mathbb{H}$. Set $\widetilde{g}_{t_n} \gamma_n (z, \overrightarrow{u}) = (z_n, \overrightarrow{u_n})$. The sequence $(d(z_n, z'))_{n \geqslant 0} = (d(\gamma_n^{-1} z_n, \gamma_n^{-1} z'))_{n \geqslant 0}$ is bounded. Moreover the sequence $(\gamma_n^{-1} z_n)_{n \geqslant 0}$ converges to $u(+\infty)$. It follows that the sequence $(\gamma_n^{-1} z')_{n \geqslant 0}$ converges to $u(+\infty)$, and hence that this point is in $L(\Gamma)$. Using Lemma 2.6, we obtain the following result

Proposition 2.7. *Let $(z, \overrightarrow{u})$ be in $T^1 \mathbb{H}$. If $\pi^1((z, \overrightarrow{u}))$ is not in $\Omega_g(T^1 S)$, then it is a divergent point.*

Clearly, when the set $\Omega_g(T^1 S)$ is compact, none of the elements of this set diverge with respect to the geodesic flow. In the general case, let us characterize the divergent points in $\Omega_g(T^1 S)$.

Proposition 2.8. *Let $(z, \overrightarrow{u})$ be in $T^1 \mathbb{H}$. Then the following are equivalent:*

(i) $\pi^1((z, \overrightarrow{u}))$ *is not positively (resp. negatively) divergent;*
(ii) $u(+\infty)$ *(resp. $u(-\infty)$) is a conical point in $L(\Gamma)$.*

Proof.

(i) $\Rightarrow$ (ii). Let $(t_n)_{n\geqslant 1}$ be an unbounded sequence in $\mathbb{R}^+$ such that $(g_{t_n}(\pi^1((z,\overrightarrow{u}))))_{n\geqslant 1}$ converges. There exists a sequence $(\gamma_n)_{n\geqslant 1}$ in Γ for which $(\gamma_n\widetilde{g}_{t_n}((z,\overrightarrow{u})))_{n\geqslant 1}$ converges to an element $(z',\overrightarrow{u}')$ in $T^1\mathbb{H}$. We have

$$\lim_{n\to+\infty} u(t_n) = u(+\infty) \quad \text{and} \quad \lim_{n\to+\infty} d(u(t_n),\gamma_n^{-1}(z')) = 0.$$

The points $u(t_n)$ belong to the ray $[z,u(+\infty))$. Furthermore

$$d(\gamma_n^{-1}(z),\gamma_n^{-1}(z')) = d(z,z'),$$

hence there exists $\varepsilon > 0$ and $N > 0$ such that

$$d(\gamma_n^{-1}(z),[z,u(+\infty))) < \varepsilon$$

whenever $n \geqslant N$. This shows that $u(+\infty)$ is conical.

(ii) $\Rightarrow$ (i). Let $(\gamma_n)_{n\geqslant 1}$ be a sequence in Γ such that

$$d(\gamma_n(z),[z,u(+\infty))) < \varepsilon.$$

It follows that there exists $s_n > 0$ satisfying $d(\gamma_n(z),u(s_n)) < \varepsilon$. Passing to a subsequence, one may assume that the sequence $(\gamma_n^{-1}\widetilde{g}_{s_n}((z,\overrightarrow{u})))_{n\geqslant 1}$ converges, which implies the convergence of the sequence $(g_{s_n}(\pi^1((z,\overrightarrow{u}))))_{n\geqslant 1}$. $\qquad\square$

We deduce from this proposition and from Proposition 2.4, the following result

Corollary 2.9. *The set $\Omega_g(T^1S)$ contains divergent points of and only if it is not compact.*

Using the preceding results and Exercise A.16 of Appendix A, we obtain the following property for semi-trajectories

Property 2.10. *Let $(z,\overrightarrow{u})$ be in $T^1\mathbb{H}$. For some $T \in \mathbb{R}$, the semi-trajectory $g_{[T,+\infty)}(\pi^1(z,\overrightarrow{u}))$ (respectively $g_{(-\infty,T]}(\pi^1(z,\overrightarrow{u}))$) is an embedding from $[T,+\infty)$ (respectively $(-\infty,T]$) into T^1S if and only if $u(+\infty)$ (respectively $u(-\infty)$) is not conical.*

2.2 Applications to geometrically finite groups

We suppose that Γ is a non-elementary, geometrically finite Fuchsian group (see Sect. I.4). Recall that there exist a Dirichlet domain $\mathcal{D}_z(\Gamma)$ and a compact subset $K \subset \mathbb{H}$, such that the intersection of this domain with the Nielsen region $N(\Gamma)$ of the group (i.e., the convex hull of the set of points in $\mathbb{H}$ belonging to geodesics with endpoints in $L(\Gamma)$) satisfies (Proposition I.4.16)

$$N(\Gamma) \cap D_z(\Gamma) = K \bigcup_{x\in L_p(\Gamma)\cap\mathcal{D}_z(\Gamma)(\infty)} H^+(x) \cap \mathcal{D}_z(\Gamma).$$

Moreover the set $L_p(\Gamma) \cap \mathcal{D}_z(\Gamma)(\infty)$ is a finite set $\{x_1, \ldots, x_n\}$, and $(H^+(x_i))_{1 \leqslant i \leqslant n}$ is a collection of horodisks centered at x_i, pairwise disjoints satisfying

$$\gamma H^+(x_i) \cap H^+(x_i) = \varnothing,$$

for all $\gamma \in \Gamma - \Gamma_{x_i}$.

Such a group Γ is also characterized by the fact that $L(\Gamma)$ is the disjoint union of the set of its conical points and that of its parabolic points (Theorem I.4.13).

Theorem 2.11. *Let Γ be a non-elementary, geometrically finite group.*

(i) *There exists a compact $K_0 \subset T^1 S$ such that, if $(z, \overrightarrow{u}) \in T^1 \mathbb{H}$ and $u(+\infty)$ (resp. $u(-\infty)$) is conical, then the set of real numbers $t > 0$ (resp. $t < 0$) for which $g_t(\pi^1((z, \overrightarrow{u}))) \in K_0$ is unbounded.*

(ii) *If $(z, \overrightarrow{u}) \in T^1 \mathbb{H}$ is such that $u(+\infty)$ (resp. $u(-\infty)$) is parabolic, then there exists $T > 0$, and a cusp of S for which the projection to S of the semi-trajectory $(g_t(\pi^1((z, \overrightarrow{u}))))_{t \geqslant T}$ (resp. $(g_t(\pi^1((z, \overrightarrow{u}))))_{t \leqslant -T}$) is included in the cusp.*
Additionally, when restricted to $\widetilde{g}_{[T,+\infty)}((z, \overrightarrow{u}))$ (resp. $\widetilde{g}_{(-\infty,-T]}((z, \overrightarrow{u}))$), the projection of $T^1 \mathbb{H}$ to $T^1 S$ is a homeomorphism onto the semi-trajectory $g_{[T,+\infty)}(\pi^1((z, \overrightarrow{u})))$ (resp. $g_{(-\infty,-T]}(\pi^1((z, \overrightarrow{u})))$).

Proof.

(i) We begin by assuming that $(z, \overrightarrow{u})$ is such that $u(+\infty)$ is conical and $u(-\infty) \in L(\Gamma)$, then $\pi^1((z, \overrightarrow{u}))$ is in $\Omega_g(T^1 S)$. To prove property (i) it is enough to prove that there exists an unbounded sequence $(t_n)_{n \geqslant 1}$ in $\mathbb{R}^+$ such that $\pi(u(t_n)) \in \pi(K)$. If this is not the case, there exists $T > 0$ such that, for all $t \geqslant T$, the point $\pi(u(t))$ is in the union of the cusps $C(x_i)$ associated to $H^+(x_i)$. Since the cusps $C(x_i)$ are disjoint, $\pi([u(T), u(+\infty)))$ is contained in a single cusp $C(x_i)$. Hence the ray $[u(T), u(+\infty))$ is in a horodisk $\gamma(H^+(x_i))$ and thus $u(+\infty) = \gamma(x_i)$, for some $\gamma \in \Gamma$, which contradicts the fact that $u(+\infty)$ is conical.

Fix $\varepsilon > 0$. Consider now an element $(z', \overrightarrow{u}')$ in $T^1 \mathbb{H}$ such that $u'(+\infty)$ is conical. Take $(z, \overrightarrow{u})$ such that $u(+\infty) = u'(+\infty)$ and $u(-\infty) \in L(\Gamma)$. The geodesic rays $[z, u'(+\infty))$ and $[z, u(+\infty))$ are asymptotic, thus there exists $T > 0$ such that $[u'(T), u'(+\infty))$ is in the ε-neighborhood of $[z, u(+\infty))$ (see Exercise 3.13). Furthermore, the preceding argument implies the existence of an unbounded sequence $(t_n)_{n \geqslant 1}$ in $\mathbb{R}^+$ such that the sequence $(g_{t_n}(\pi^1((z, \overrightarrow{u}))))_{n \geqslant 1}$ is in $\pi(K)$. From these properties one deduces the existence of an unbounded sequence $(t'_n)_{n \geqslant 1} \subset \mathbb{R}^+$, such that the sequence $(\pi((z', \overrightarrow{u}')(t'_n)))_{n \geqslant 1}$ is in the ε-neighborhood of $\pi(K)$. Such a neighborhood is compact.

(ii) To avoid unnecessary notation, we present an argument on S which can be extended $T^1 S$. Let $(z, \overrightarrow{u}) \in T^1 \mathbb{H}$ such that $u(+\infty)$ is parabolic. One can suppose $u(+\infty) = x_i$ (Corollary I.4.10). The projection q of the quotient

$\Gamma_{x_i} \backslash H^+(x_i)$ to the cusp $C(x_i)$ is a homeomorphism (see Sect. I.3.22). For large enough T, the ray $[u(T), x_i)$ is contained in $H^+(x_i)$. Furthermore, if one restricts the projection of $H^+(x_i)$ to $\Gamma_{x_i} \backslash H^+(x_i)$ to this ray, the resulting map p is a homeomorphism. Thus $q \circ p$ is a homeomorphism of $[u(T), x_i)$ onto $\pi([u(T), x_i))$.   □

Let E_i denote the set of all $(z, \overrightarrow{u})$ in $T^1\mathbb{H}$ such that z is in the horocycle $H(x_i)$ associated to $H^+(x_i)$. Notice that $\pi^1(E_i)$ is a compact subset of $T^1 S$. Take $(z', \overrightarrow{u'})$ in $T^1\mathbb{H}$ whose positive endpoint $u'(+\infty)$ is parabolic; there exist $i \in \{1, \ldots, n\}$ and γ in Γ such that $\gamma(v(+\infty)) = x_i$. A geodesic having x_i as an endpoint intersects $H(x_i)$, hence $\gamma \widetilde{g}_{\mathbb{R}}((z', \overrightarrow{u'})) \cap E_i \neq \varnothing$. It follows that the union of the compact set K_0 given by Theorem 2.11(i) with the projection to $T^1 S$ of the E_i is a compact set intersected by all the semi-trajectories $g_{\mathbb{R}^+}(\pi^1((z, \overrightarrow{u})))$, with $u(+\infty) \in L(\Gamma)$.

Corollary 2.12. *Let Γ be a non-elementary, geometrically finite group. There exists a compact subset of $T^1 S$ intersected by every semi-trajectory $g_{\mathbb{R}}^+(\pi^1((z, \overrightarrow{u})))$, with $u(+\infty) \in L(\Gamma)$.*

Some of the trajectories of $g_{\mathbb{R}}$ on $\Omega_g(T^1 S)$ are closed. This is the case if, for example, $u(-\infty)$ and $u(+\infty)$ are parabolic (Theorem 2.11). Some trajectories are compact. In the following section it will be shown that the compactness corresponds to the case in which the points $u(+\infty)$ and $u(-\infty)$ are fixed by a hyperbolic isometry of Γ. Some trajectories are dense (see Sect. 4). Others are very "chaotic." Having information about the conical and parabolic nature of $u(-\infty)$ and $u(+\infty)$ alone is not sufficient to generally describe the topology of $g_{\mathbb{R}}(\pi^1((z, \overrightarrow{u})))$. In Chap. IV, we will study the case in which Γ is a Schottky group generated by two hyperbolic isometries. Using doubly-infinite sequences as a coding of $\widetilde{\Omega}_g(T^1 S)$, we will establish a correspondence between the topological dynamics of $g_{\mathbb{R}}$ on $\Omega_g(T^1 S)$ and that of the shift on the space of these doubly-infinite sequences. We will then see the emergence of a wide variety of topological structures for trajectories of $g_{\mathbb{R}}$. A notable example of this is the existence of minimal compact sets which are $g_{\mathbb{R}}$-invariant, yet are not periodic trajectories.

When $\Omega_g(T^1 S)$ is not compact, this set contains unbounded trajectories. For example, if $u(+\infty)$ or $u(-\infty)$ is parabolic, then $(g_t(\pi^1(u))_{t \in \mathbb{R}}$ is unbounded. This condition on $u(+\infty)$ and $u(-\infty)$ is sufficient but not necessary. In Chap. VII, in the context of the modular surface, we will relate the boundedness of $g_{\mathbb{R}}(\pi^1(u))$ to a property of the continued fraction expansion of $u(-\infty)$ and $u(+\infty)$ (Theorem VII.3.4).

3 Periodic trajectories and their periods

Returning to the general case of a non-elementary Fuchsian group Γ, we focus on the periodic trajectories of the geodesic flow on $T^1 S$. Recall that

$\pi^1((z, \overrightarrow{u})) \in T^1S$ is periodic for the geodesic flow if there exists some $t > 0$ such that $g_t(\pi^1((z, \overrightarrow{u}))) = \pi^1((z, \overrightarrow{u}))$.

3.1 Density of periodic trajectories

The following proposition gives a characterization of the periodic elements.

Proposition 3.1. *Let $(z, \overrightarrow{u})$ be in $T^1\mathbb{H}$. The following are equivalent:*

(i) *the element $\pi^1((z, \overrightarrow{u}))$ is periodic;*
(ii) *there exists a hyperbolic isometry γ in Γ such that $u(+\infty) = \gamma^+$ and $u(-\infty) = \gamma^-$.*

Proof.
   (ii) $\Rightarrow$ (i). The isometry γ leaves the oriented geodesic $(\gamma^- \gamma^+)$ invariant. Furthermore, Property I.2.8 implies that given a point z on this geodesic, one has that $d(z, \gamma(z)) = \ell(\gamma)$. Therefore $\gamma((z, \overrightarrow{u})) = \widetilde{g}_{\ell(\gamma)}((z, \overrightarrow{u}))$.
   (i) $\Rightarrow$ (ii). There exists $t > 0$ and γ in Γ such that $\widetilde{g}_t((z, \overrightarrow{u})) = \gamma((z, \overrightarrow{u}))$. It follows that $\widetilde{g}_{nt}((z, \overrightarrow{u})) = \gamma^n((z, \overrightarrow{u}))$. Thus γ fixes $u(+\infty)$ and $u(-\infty)$. These two points are distinct and γ is not the identity, hence this isometry is hyperbolic.   □

Let $\pi^1((z, \overrightarrow{u}))$ in T^1S be a periodic point for the geodesic flow. Applying Proposition 3.1, we obtain a hyperbolic isometry $\gamma \in \Gamma$ fixing the points $u(+\infty)$ and $u(-\infty)$ such that $t = d(z, \gamma(z))$. Since Γ is discrete, the subgroup of hyperbolic isometries fixing $u(+\infty)$ and $u(-\infty)$ is generated by one *primitive* element $\gamma_0 \neq \mathrm{Id}$ (i.e., there is no isometry h in Γ satisfying $h^n = \gamma_0$ for $n > 1$). It follows that the set of real numbers t such that $g_t(\pi^1((z, \overrightarrow{u}))) = \pi^1((z, \overrightarrow{u}))$, which is a closed subgroup of $(\mathbb{R}, +)$, is the set $\mathbb{Z}\ell(\gamma_0)$. Since $\ell(\gamma_0)$ is the smallest $t > 0$ such that $g_t(\pi^1((z, \overrightarrow{u}))) = \pi^1((z, \overrightarrow{u}))$, it is called the *period* of $\pi^1((z, \overrightarrow{u}))$ and is denoted T_u.

Let $\mathrm{Hyp}(\Gamma)$ denote the set of conjugacy classes in Γ of primitive hyperbolic isometries of Γ. According to Proposition 3.1, there is a bijection from $\mathrm{Hyp}(\Gamma)$ onto the set of periodic trajectories of $g_\mathbb{R}$, which sends an equivalence class $[\gamma]$ to the trajectory $g_\mathbb{R}(\pi^1((z, \overrightarrow{u})))$, where $(z, \overrightarrow{u})$ is an element of $T^1\mathbb{H}$ satisfying $u(+\infty) = \gamma^+$ and $u(-\infty) = \gamma^-$. Since the group Γ is not elementary, by Corollary II.1.3, the set $\mathrm{Hyp}(\Gamma)$ is infinite. This allows us to state the following property.

Property 3.2. *The set $\Omega_g(T^1S)$ contains infinitely many periodic trajectories.*

In Chap. IV, we will give another proof of the following Theorem using the techniques of symbolic dynamics in the particular case of convex-cocompact Schottky groups (see Sect. IV.2.1).

Theorem 3.3. *The set of periodic elements is dense in $\Omega_g(T^1S)$.*

Proof. Fix $(z, \vec{u})$ in $\widetilde{\Omega}_g(T^1 S)$ (i.e., $u(+\infty)$ and $u(-\infty)$ belong to $L(\Gamma)$). Proposition 1.6 reduces the proof of this theorem to proving that there exists a sequence of hyperbolic isometries $(\gamma_n)_{n \geqslant 1}$ in Γ, which satisfy $\lim_{n \to +\infty} \gamma_n^+ = u(+\infty)$ and $\lim_{n \to +\infty} \gamma_n^- = u(-\infty)$.

By Lemma 2.2, there exists $(\gamma_n)_{n \geqslant 1}$ in Γ such that $\lim_{n \to +\infty} \gamma_n(i) = u(+\infty)$ and $\lim_{n \to +\infty} \gamma_n^{-1}(i) = u(-\infty)$.

We will now show that for large enough n, γ_n is hyperbolic. For this task, we work in the Poincaré disk with a center 0. Let $D_0(\gamma_n)$ be the closed half-plane bounded by the perpendicular bisector of the segment $[0, \gamma_n(0)]_h$, containing $\gamma_n(0)$. The sequence of Euclidean diameters of $D_0(\gamma_n^{\pm 1})$ converges to zero since $\lim_{n \to +\infty} \gamma_n^{\pm 1}(0) = u(+\infty)$. The points $u(-\infty)$ and $u(+\infty)$ are distinct, thus for large enough n, the half diks $D_0(\gamma_n)$ and $D_0(\gamma_n^{-1})$ are disjoint. Property I.2.7, implies that γ_n is hyperbolic. Moreover, γ_n^+ (respectively γ_n^-) is in the boundary at infinity of $D_0(\gamma_n)$ (respectively $D_0(\gamma_n^{-1})$), hence $\lim_{n \to +\infty} \gamma_n^+ = u(+\infty)$ and $\lim_{n \to +\infty} \gamma_n^- = u(-\infty)$. $\qquad\square$

3.2 Length spectrum

We define a *geodesic* (respectively *geodesic segment*) of the surface S as the image under the canonical projection π of a geodesic (respectively geodesic segment) of $\mathbb{H}$ (Fig. III.8).

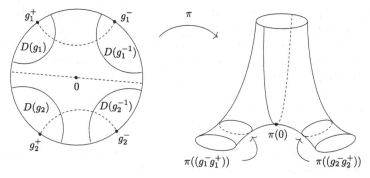

Fig. III.8. $\Gamma = S(g_1, g_2)$, g_1 and g_2 hyperbolic

If the group Γ does not contain any elliptic elements, S inherits its Riemannian structure. Geodesics on S coincide with geodesics for this Riemannian structure.

Exercise 3.4. Prove that a geodesic of S is compact if and only if it is the projection of a periodic trajectory of $g_{\mathbb{R}}$ to S.

Let γ be a hyperbolic element of Γ. Consider a primitive element h in Γ having the same axis as γ. If the group Γ does not contain any elliptic element, then S is a Riemannian surface and the real number $\ell(h)$ is the *length*, in the Riemannian sense, of the geodesic $\pi(\gamma^-\gamma^+)$. More generally, one defines the *length* of $\pi(\gamma^-\gamma^+)$, as

$$\text{length}_S(\pi(\gamma^-\gamma^+)) = \ell(h).$$

Notice that, if $(z, \overrightarrow{u})$ is such that $u(+\infty) = \gamma^+$ and $u(-\infty) = \gamma^-$, then $\pi^1((z, \overrightarrow{u}))$ is periodic and its period T_u is given by:

$$T_u = \text{length}_S(\pi(\gamma^-\gamma^+)).$$

It follows that the set of lengths of compact geodesics of S is in one-to-one correspondence with the set $SP(g_\mathbb{R})$ of periods associated to the periodic trajectories of $g_\mathbb{R}$.

We are interested in the set of all nT, where $n \in \mathbb{N}$ and $T \in SP(g_\mathbb{R})$. It is in one-to-one correspondence with the set $\mathcal{L}(\Gamma) = \{\ell(\gamma) \mid \gamma \in \Gamma\}$, called the *length spectrum* of Γ.

Property 3.5. *Let Γ be a geometrically finite group and $(\gamma_n)_{n \geqslant 1}$ be a sequence of hyperbolic isometries in Γ. If $(\ell(\gamma_n))_{n \geqslant 1}$ is a bounded sequence, then there exist k isometries $g_1, \ldots, g_k$ in Γ such that, for all $n \geqslant 1$, the isometry γ_n is conjugate in Γ to one of the g_i.*

Proof. Let $(z_n, \overrightarrow{u_n})$ be in $T^1\mathbb{H}$ such that $u_n(+\infty) = \gamma_n^+$ and $u_n(-\infty) = \gamma_n^-$. Since the group Γ is geometrically finite, Corollary 2.11 implies that some compact subset of T^1S is intersected by all trajectories $g_\mathbb{R}(\pi^1((z_n, \overrightarrow{u_n})))$. It follows that after conjugating γ_n, replacing $(z_n, \overrightarrow{u_n})$ with an element of $\tilde{g}_\mathbb{R}((z_n, \overrightarrow{u_n}))$, and passing to a subsequence, one may assume that the sequence $((z_n, \overrightarrow{u_n}))_{n \geqslant 1}$ converges to $(z', \overrightarrow{u}')$. One has $d(\gamma_n(z_n), z_n) = \ell(\gamma_n)$. Furthermore, passing to another subsequence, $(\ell(\gamma_n))_{n \geqslant 1}$ converges. Hence there exists $\varepsilon > 0$ and $M > 0$ such that $d(\gamma_n(z'), z') \leqslant \varepsilon$ for all $n \geqslant M$. Since the group Γ is discrete, the set of such γ_n is finite. $\qquad\square$

In terms of periods of the geodesic flow, the preceding property implies that, if a sequence $(T_{u_n})_{n \geqslant 1}$ in $SP(g_\mathbb{R})$ is bounded then the elements $\pi^1(u_n)$ are contained in a finite number of periodic trajectories.

Another consequence is that for all $t > 0$, the subset of $\text{Hyp}(\Gamma)$ composed of the classes $[\gamma]$ satisfying $\ell(\gamma) \leqslant t$, is finite.

In particular, there is a finite number of compact geodesics on S whose length is less than that of all other geodesics. On the modular surface, applying Corollary II.4.14, we obtain that there is only one such geodesic and that its length is equal to $4\ln\mathcal{N}$, where $\mathcal{N}$ is the golden ratio.

Theorem 3.6. *The additive group generated by the length spectrum $\mathcal{L}(\Gamma)$ (and thus by the periods $SP(g_\mathbb{R})$) is dense in $\mathbb{R}$.*

Proof. Since the group Γ is non-elementary, Corollary II.1.3 implies that it contains a Schottky group generated by a pair of hyperbolic isometries h and g. Let us use the Poincaré disk model and fix a point 0. One may assume that the half-planes $D_0(h) = D(h), D_0(h^{-1}) = D(h^{-1}), D_0(g) = D(g), D_0(g^{-1}) = D(g^{-1})$, defined in Sect. I.2.2, are located as in Fig. III.9.

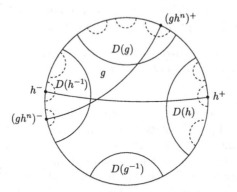

Fig. III.9.

By Proposition II.1.4, the geodesics $(h^- h^+)$ and $((gh^n)^- (gh^n)^+)$ intersect each other at a point z_n in $\mathbb{D}$ for all $n > 0$. The sequence $(z_n)_{n \geqslant 1}$ converges to h^-. Thus one may assume that the points z_n are all distinct. One has

$$\ell(gh^n) = d(z_n, gh^n(z_n)) \quad \text{and} \quad \ell(h) = d(z_n, h(z_n)).$$

Furthermore, z_n is not in the axis of gh^{n+1} thus from Property I.2.7(i), one has $\ell(gh^{n+1}) < d(z_n, gh^{n+1}(z_n))$. It follows that for all $n \geqslant 1$

$$\ell(gh^{n+1}) < d(h^{-n}g^{-1}(z_n), h(z_n)),$$

thus

(∗) $$\ell(gh^{n+1}) < \ell(gh^n) + \ell(h).$$

Let us return to the Poincaré half-plane and choose an isometry $\gamma(z) = (az + b)/(cz + d)$ in $\mathbb{H}$. For such an isometry, we have

$$\cosh\left(\frac{\ell(\gamma)}{2}\right) = \frac{|a + d|}{2},$$

where cosh denotes the hyperbolic cosine.

Using this relationship and supposing, after conjugating Γ, that $h(z) = \lambda z$ with $\lambda > 1$, one obtains

$$\lim_{n \to +\infty} \ell(gh^{n+1}) - \ell(gh^n) = \ell(h).$$

If $\mathcal{L}(\Gamma)$ generates a discrete group, then $\ell(gh^{n+1}) - \ell(gh^n) = \ell(h)$ for large enough n, which contradicts the inequality (∗). □

4 Dense trajectories

In this section, we prove the existence of trajectories of the geodesic flow that are dense in its non-wandering set.

We use the criterion (Proposition 1.6) relating the action of $g_{\mathbb{R}}$ on T^1S to the dual action of Γ on the set of oriented geodesics $\tilde{g}_{\mathbb{R}}\backslash T^1\mathbb{H}$, or more precisely on the set $L(\Gamma) \times^{\Delta} L(\Gamma)$ defined to be the product $L(\Gamma) \times L(\Gamma)$ minus its diagonal. Let us begin by proving the following lemma.

Lemma 4.1. *Let Γ be a non-elementary Fuchsian group. For any open, non-empty subsets O and V of $L(\Gamma) \times^{\Delta} L(\Gamma)$, there exists $\gamma \in \Gamma$ such that $\gamma(O) \cap V \neq \varnothing$.*

Proof. One can assume that O and V are products of open, non-empty sets $O = O_1 \times O_2$, $V = V_1 \times V_2$. Since the set $L(\Gamma)$ is minimal, V_1 contains the attractive fixed point γ^+ of a hyperbolic isometry γ in Γ. It follows that, for large enough n, $\gamma^n O_1 \cap V_1 \neq \varnothing$. Furthermore, from Theorem 3.3, there exists a hyperbolic isometry h in Γ such that h^- is contained in $\gamma^n O_1 \cap V_1$ and h^+ is in V_2. Thus for large enough k, one has $h^k\gamma^n O_2 \cap V_2 \neq \varnothing$. Moreover, since the point h^- is in $\gamma^n O_1 \cap V_1$, we have $h^k\gamma^n O_1 \cap V_1 \neq \varnothing$. This implies that $h^k\gamma^n O \cap V \neq \varnothing$. $\square$

Theorem 4.2. *There exists $(z, \vec{u})$ in $\Omega_g(T^1S)$ such that $\overline{g_{\mathbb{R}}(\pi^1((z, \vec{u})))} = \Omega_g(T^1S)$.*

Proof. Consider a countable family of open, non-empty subsets $(O_n)_{n \geqslant 1}$ in $L(\Gamma) \times^{\Delta} L(\Gamma)$ such that every open subset of $L(\Gamma) \times^{\Delta} L(\Gamma)$ contains one of the O_n. Fix an open, non-empty subset O of $L(\Gamma) \times^{\Delta} L(\Gamma)$. After Lemma 4.1, there exists γ_1 in Γ such that $\gamma_1 O \cap O_1 \neq \varnothing$. Let K_1 be an open, relatively compact subset of O such that $\gamma_1\overline{K}_1$ is in O_1. Repeating this argument and replacing O with K_1 and O_1 with O_2, one obtains γ_2 in Γ and an open, relatively compact subset K_2 such that $\overline{K}_2 \subset K_1$ and $\gamma_2\overline{K}_2 \subset O_2$. Continuing this process, one obtains a sequence $(K_n)_{n \geqslant 1}$ of open, relatively compact, nested subsets. The set $\bigcap_{n=1}^{+\infty} \overline{K}_n$ is not empty. Let x in $\bigcap_{n=1}^{+\infty} \overline{K}_n$. For all $n \geqslant 1$, the point $\gamma_n(x)$ is in O_n. Consider a point x' in $L(\Gamma) \times^{\Delta} L(\Gamma)$ and a neighborhood V' of x'. This neighborhood contains an open set O_n, thus $\gamma_n(x)$ is in V'. The orbit Γx therefore intersects every neighborhood of x', which shows that x' is in $\overline{\Gamma x}$. It follows that $\overline{\Gamma x} = L(\Gamma) \times^{\Delta} L(\Gamma)$. It is sufficient to apply Proposition 1.6 to complete the proof. $\square$

Theorem 4.2 will be proved again in the next chapter for the particular case of convex-cocompact Schottky groups. This case provides a characterization of dense trajectories $g_{\mathbb{R}}(\pi^1((z, \vec{u})))$ in terms of the coding of $u(-\infty)$ and $u(+\infty)$ (see Sect. IV.2.2).

The following theorem implies Lemma 4.1. It will be proved later using the horocycle flow (see Sect. V.3).

Theorem 4.3 (Topological mixing). *Let O and V be two open, non-empty subsets of $\Omega_g(T^1S)$. There exists $T > 0$ such that for all $t \geqslant T$, $g_t O \cap V \neq \varnothing$.*

5 Comments

The geodesic flow is also well-defined on the unitary tangent bundle of a complete Riemannian manifold and in particular on the unitary tangent bundle of a pinched Hadamard manifold T^1X, and on its quotient by a torsion-free Kleinian group Γ [14, 49].

The results of this chapter, with the exception of Property 3.6 and Theorem 4.3 and their proofs, are inspired by two articles on the geodesic flow on $\Gamma\backslash T^1X$ by P. Eberlein ([26] and [28]).

In this general context, the length spectrum is a source of open problems. One of these problems consists of knowing whether or not Property 3.6 about the density of the length spectrum always holds for Schottky groups. It is open if X is not a symmetric space and if $\dim(X) \geqslant 3$ [11, 18, 29]. The density property of the length spectrum is especially important since it is equivalent to that of the topological mixing of the geodesic flow [18].

In this book we do not study the metric properties of the geodesic flow. Let us give the reader some idea of these properties. Consider a Fuchsian group Γ acting on the Poincaré disk $\mathbb{D}$. A construction due to D. Sullivan [61, 62] allows us to obtain from a Patterson measure m on $L(\Gamma)$ (introduced in the Comments at the end of Chap. I) a measure M on $L(\Gamma) \times^\Delta L(\Gamma) \times \mathbb{R}$ defined as

$$M(dx\,dy\,ds) = \frac{m(dx)m(dy)\,ds}{|x-y|^{2\delta(\Gamma)}},$$

where $\delta(\Gamma)$ is the critical exponent of the Poincaré series associated with Γ.

Identifying $T^1\mathbb{D}$ with triplets of points (x, y, s) such that $x \neq y$ and $s \in \mathbb{R}$, the measure M becomes a Γ-invariant measure on $T^1\mathbb{D}$. This measure is also flow invariant with respect to $\widetilde{g}_\mathbb{R}$. Therefore it induces a measure $\overline{M}$ on $\Gamma\backslash T^1\mathbb{D}$ which is again $g_\mathbb{R}$-flow invariant and preserves the non-wandering set of this flow [48]. If Γ is a lattice, $\overline{M}$ is proportional to the Liouville measure. More generally, if Γ is geometrically finite, $\overline{M}$ is finite and the geodesic flow is ergodic and mixing [5, 54].

This construction can be generalized to the case of pinched Hadamard manifolds [11], but the geometric finiteness of Γ does not necessarily imply the finiteness of $\overline{M}$ [23, 54].

The measure $\overline{M}$ plays a crucial role in solving counting problems, like for example counting points in an orbit or counting closed geodesics.

IV

Schottky groups and symbolic dynamics

Throughout this chapter, the group Γ will designate a Schottky group generated by two hyperbolic isometries g_1, g_2 (see Sect. II.1). By definition, such a group admits a Dirichlet domain centered at a point designated to be 0 in the Poincaré disk. The possible cases are diagrammed below in Fig. IV.1. For further details, the reader may refer to Sect. II.1.

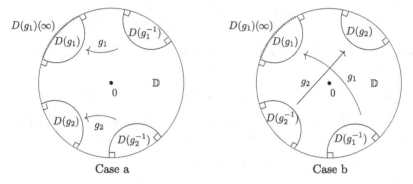

Case a   Case b

Fig. IV.1.

Recall that if g is an isometry which does not fix 0, then the set $D_0(g)$ represents the closed half-plane in the disk $\mathbb{D}$, containing the point $g(0)$, and bounded by the perpendicular bisector of the segment $[0, g(0)]_h$.

The surface $S = \Gamma \backslash \mathbb{D}$ is homeomorphic to a sphere with three points removed, or to a torus minus one point. In each of these cases, the limit set of Γ is composed exclusively of conical points (Property II.1.13). Hence the non-wandering set of the geodesic flow, $\Omega_g(T^1 S)$, is compact (Proposition III.2.4).

The goal of this chapter is to encode the trajectories of the geodesic flow restricted to $\Omega_g(T^1 S)$ into doubly-infinite sequences, and to develop this point of view into a method of studying the dynamics of this flow. This symbolic approach will allow us to present new proofs of Theorems III.3.3 and III.4.2.

F. Dal'Bo, *Geodesic and Horocyclic Trajectories*, Universitext,
DOI 10.1007/978-0-85729-073-1_4, © Springer-Verlag London Limited 2011

Moreover we will complete the latter theorem by characterizing the dense trajectories of $\Omega_g(T^1S)$ in terms of sequences. As applications, we will construct, in the general case of a non-elementary Fuchsian group Γ', trajectories of the geodesic flow on $\Omega_g(\Gamma' \backslash T^1\mathbb{D})$ which are neither periodic nor dense.

1 Coding

Recall that Σ^+ represents the set of sequences $s = (s_i)_{i \geqslant 1}$ satisfying

$$s_i \in \mathcal{A} = \{g_1^{\pm 1}, g_2^{\pm 1}\} \quad \text{and} \quad s_{i+1} \neq s_i^{-1}.$$

We have also defined $f : \Sigma^+ \to L(\Gamma)$ to be the map which sends a sequence $s = (s_i)_{i \geqslant 1}$ to the following point

$$f(s) = \lim_{n \to +\infty} s_1 \cdots s_n(0) \quad \text{(see Sects. II.1 and II.2).}$$

Since g_1 and g_2 are hyperbolic, this map is a bijection (Proposition II.2.2). We equip Σ^+ with the following metric δ:

$$\delta(s, s') = \begin{cases} 0 & \text{if } s = s', \\ \dfrac{1}{\min\{i \geqslant 1 \mid s_i \neq s_i'\}} & \text{otherwise.} \end{cases}$$

Exercise 1.1. Prove that δ is a metric, and that the metric space (Σ^+, δ) is compact.

Lemma 1.2. *The map* $f : (\Sigma^+, \delta) \to L(\Gamma)$ *is a homeomorphism.*

Proof. It is sufficient to prove that this map is continuous. Consider a sequence $(u_n)_{n \geqslant 1}$ in Σ^+ which converges to an element s of Σ^+. Define

$$u_n = (u_{n,i})_{i \geqslant 1} \quad \text{and} \quad s = (s_i)_{i \geqslant 1}.$$

For all $k \geqslant 2$, there exists $N > 1$ such that, for each $1 \leqslant i \leqslant k$ and $n \geqslant N$, one has $u_{n,i} = s_i$. Let T be the shift operator on Σ^+. For all $n \geqslant N$, we have

$$f(u_n) = s_1 \cdots s_{k-1}(f(T^{k-1}(u_n))).$$

Since the point $f(T^{k-1}(u_n))$ belongs to the boundary at infinity of the half-disk $D_0(s_k)$, the point $f(u_n)$ is in $D(s_1, \ldots, s_k)(\infty) = s_1 \cdots s_{k-1} D_0(s_k)(\infty)$. This set also contains $f(s)$ since

$$f(s) = s_1 \cdots s_k \lim_{n \to +\infty} s_{k+1} \cdots s_n(0).$$

According to Lemma II.1.10, the sequence of Euclidean diameters of the nested sets $D(s_1, \ldots, s_k)$ converges to 0 as k tends to $+\infty$. Therefore $\lim_{n \to +\infty} |f(u_n) - f(s)| = 0$. $\square$

We now consider the set Σ of doubly-infinite sequences $S = (S_i)_{i \in \mathbb{Z}}$ which satisfy the following conditions:

$$(S_i)_{i \geqslant 1} \in \Sigma^+, \quad (S_{-i+1}^{-1})_{i \geqslant 1} \in \Sigma^+ \quad \text{and} \quad S_1 \neq S_0^{-1}.$$

We associate to S be in Σ, two sequences defined by

$$S^+ = (S_i)_{i \geqslant 1} \quad \text{and} \quad S^- = (S_{-i+1}^{-1})_{i \geqslant 1}.$$

The condition $S_1 \neq S_0^{-1}$ together with Property II.1.4(i) implies that the points $f(S^+)$ and $f(S^-)$ are distinct.

The distance function δ on Σ^+ induces a distance function Δ on Σ defined by

$$\Delta(S, S') = \sqrt{\delta^2(S^+, S'^+) + \delta^2(S^-, S'^-)}.$$

We still denote by $T : \Sigma \to \Sigma$ the shift operator $T(S) = (S_{i+1})_{i \in \mathbb{Z}}$. This operator is a bijection of Σ onto itself.

Exercise 1.3. Prove that the shift operator T on (Σ, Δ) is continuous.

Exercise 1.4. Prove that the metric space (Σ, Δ) is compact.

We will establish a correspondence between the topology of orbits of T on (Σ, Δ) and the trajectories of the geodesic flow on $\Omega_g(T^1 S)$. To do this, recall first that $L(\Gamma) \times^\Delta L(\Gamma)$ denotes the product of $L(\Gamma)$ with itself minus its diagonal, and denote by $F : \Sigma \to L(\Gamma) \times^\Delta L(\Gamma)$, the map defined by

$$F(S) = (x(S^-), x(S^+)).$$

This map is continuous and injective since the map f is. However, it is not surjective since $x(S^-)$ and $x(S^+)$ respectively belong to the disjoint arcs $D_0(S_0^{-1})(\infty)$ and $D_0(S_1)(\infty)$.

Lemma 1.5. *Given (x_-, x_+) in $L(\Gamma) \times^\Delta L(\Gamma)$, there exist γ in Γ and S in Σ such that $\gamma(x_-, x_+) = F(S)$.*

Proof. Let $a = (a_i)_{i \geqslant 1}$ and $b = (b_i)_{i \geqslant 1}$ be the elements of Σ^+ such that $x_- = f(a)$ and $x_+ = f(b)$. By hypothesis $x_- \neq x_+$. Consider the smallest $N \geqslant 1$ for which $a_N \neq b_N$. Let S denote the doubly-infinite sequence defined by

$$S_i = \begin{cases} a_{N+i-1} & \text{if } i \geqslant 1, \\ b_{N-i}^{-1} & \text{if } i \leqslant 0. \end{cases}$$

This sequence belongs to Σ. If $N = 1$, then $F(S) = (x_-, x_+)$; otherwise $F(S) = (a_1^{-1} \cdots a_{N-1}^{-1}(x_-), a_1^{-1} \cdots a_{N-1}^{-1}(x_+))$. $\qquad \square$

This lemma shows that the map F is a surjection to a set of representatives of Γ-orbits on $L(\Gamma) \times^\Delta L(\Gamma)$.

Recall that π^1 denotes the projection of $T^1 \mathbb{H}$ to $T^1 S$.

Proposition 1.6. *Let S, S' be in Σ and $(z, \overrightarrow{u}), (z', \overrightarrow{u}')$ in $T^1\mathbb{D}$ such that*

$$(u(-\infty), u(+\infty)) = F(S), \quad (u'(-\infty), u'(+\infty)) = F(S').$$

Then the following are equivalent:

(i) $S' \in \overline{T^{\mathbb{Z}}(S)}$;
(ii) $\pi^1((z', \overrightarrow{u}')) \in \overline{g_{\mathbb{R}}(\pi^1((z, \overrightarrow{u})))}$.

Proof. Part (ii) is equivalent to the existence of sequences $(s_n)_{n \geqslant 1}$ in $\mathbb{R}$ and $(\gamma_n)_{n \geqslant 1}$ in Γ such that $\lim_{n \to +\infty} \gamma_n(\widetilde{g}_{s_n}(u)) = u'$. According to Proposition III.1.6, this is in turn equivalent to the existence of a sequence $(\gamma_n)_{n \geqslant 1}$ in Γ satisfying

(iii) $\qquad \lim_{n \to +\infty} (\gamma_n(u(-\infty)), \gamma_n(u(+\infty))) = (u'(-\infty), u'(+\infty)).$

The implication (i) $\Rightarrow$ (iii) follows directly from the equality

$$F(T^n(S)) = (\gamma_n(u(-\infty)), \gamma_n(u(+\infty))),$$

where $\gamma_n = S_n^{-1} \cdots S_1^{-1}$ if $n > 0$ and $\gamma_n = S_{n+1} \cdots S_0$ otherwise.

Let us show (iii) $\Rightarrow$ (i). Consider a sequence $(\gamma_n)_{n \geqslant 1}$ in Γ satisfying

$$\lim_{n \to +\infty} (\gamma_n(u(-\infty)), \gamma_n(u(+\infty))) = (u'(-\infty), u'(+\infty)).$$

Write γ_n in the form of a reduced word $\gamma_n = a_{n,1} \cdots a_{n,\ell_n}$. If there exists a subsequence $(\gamma_{n_k})_{k \geqslant 1}$ satisfying

$$\gamma_{n_k} = S_{\ell_{n_k}}^{-1} \cdots S_1^{-1} \quad \text{or} \quad \gamma_{n_k} = S_{-\ell_{n_k}+1} \cdots S_0,$$

then $(\gamma_{n_k}(u(-\infty)), \gamma_{n_k}(u(+\infty))) = F(T^{\ell_{n_k}}(S))$. Since F is a homeomorphism onto its image, it follows that $\lim_{n \to +\infty} T^{\ell_{n_k}}(S) = S'$.

Otherwise, for large enough n, γ_n is distinct from both $S_{\ell_n}^{-1} \cdots S_1^{-1}$ and $S_{-\ell_n+1} \cdots S_0$. Thus Property II.1.4 implies that the points $\gamma_n(u(-\infty))$ and $\gamma_n(u(+\infty))$ are in the same circular arc of $D(a_{n,1})(\infty)$. Passing to a subsequence, one may assume that $a_{n,1} = a_1$. Thus the points $u'(-\infty)$ and $u'(+\infty)$ are elements of $D(a_1)(\infty)$. This is impossible since by hypothesis $u'(-\infty) \in D(S_0'^{-1})(\infty)$, $u'(+\infty) \in D(S_1')(\infty)$ and $S_0'^{-1} \neq S_1'$. $\qquad \square$

2 The density of periodic and dense trajectories

2.1 An alternate proof of Theorem III.3.3

We first establish relationship between the sequences in Σ which are periodic with respect to the shift T, and the elements of $\Omega_g(T^1S)$ which are periodic with respect to the geodesic flow.

Let $S = (S_i)_{i \geqslant 1}$ in Σ be T-periodic of period n. The point $f(S^+)$ is the attracting fixed point of $\gamma = S_1 \cdots S_n$, and $f(S^-)$ is the repulsive one. If $(z, \overrightarrow{u}) \in T^1 \mathbb{D}$ satisfies $(u(-\infty), u(+\infty)) = F(S)$, then by Proposition III.3.1, the element $\pi^1((z, \overrightarrow{u}))$ is periodic with respect to the geodesic flow.

Conversely, if $\pi^1((z, \overrightarrow{u}))$ is periodic with respect to $g_{\mathbb{R}}$, after replacing $(z, \overrightarrow{u})$ with an element of $\Gamma(z, \overrightarrow{u})$, one can assume that there exists a primitive hyperbolic isometry $\gamma = a_1 a_2 \cdots a_n$ with $a_i \in \mathcal{A}$, $a_{i+1} \neq a_i^{-1}$ and $a_1 \neq a_n^{-1}$ satisfying $u(-\infty) = \gamma^-$ and $u(+\infty) = \gamma^+$. Since $a_1 \neq a_n^{-1}$, the sequences $f^{-1}(\gamma^+)$ and $f^{-1}(\gamma^-)$ are T-periodic of period n. Consider the doubly-infinite periodic sequence $(S_i)_{i \in \mathbb{Z}}$ of period n defined by $S_1 = a_1, \ldots, S_n = a_n$. This sequence belongs to Σ and satisfies $F(S) = (\gamma^-, \gamma^+)$. In conclusion we obtain the following property:

Property 2.1. Let $(z, \overrightarrow{u})$ be in $T^1 \mathbb{D}$. The element $\pi^1((z, \overrightarrow{u}))$ is $g_{\mathbb{R}}$-periodic if and only if there exists S in Σ which is T-periodic, and an isometry γ in Γ such that
$$\gamma(u(-\infty), u(+\infty)) = F(S).$$

Using this dictionary between periodic sequences and periodic trajectories for the geodesic flow, we give another proof of the density of the set of the $g_{\mathbb{R}}$-periodic elements in $\Omega_g(\Gamma \backslash T^1 \mathbb{D})$ (Theorem III.3.3), when Γ is a Schottky group.

A proof of Theorem III.3.3 using symbolic dynamics

Let $\pi^1((z, \overrightarrow{u}))$ be in $\Omega_g(\Gamma \backslash T^1 \mathbb{H})$. Lemma 1.5 implies that there exist γ in Γ, and S in Σ such that $\gamma(u(-\infty), u(+\infty)) = F(S)$. For each $n \geqslant 1$, choose a_{n+1} in $\mathcal{A} - \{S_n^{-1}, S_{-n}^{-1}\}$. Consider the sequence $(U_k)_{k \geqslant 1}$ of elements of Σ in which each term $U_k = (U_{k,i})_{k \geqslant 1}$ is a periodic sequence of period $2k + 2$ such that

$$U_{k,1} = S_1, \qquad U_{k,2} = S_2, \ldots, U_{k,k} = S_k,$$
$$U_{k,k+1} = a_{k+1},$$
$$U_{k,k+2} = S_{-k}, \qquad U_{k,k+3} = S_{-k+1}, \ldots, U_{k,2k+2} = S_0.$$

Then $\Delta(U_k, S) \leqslant \sqrt{2}/k$, which further implies that $\lim_{k \to +\infty} U_k = S$. For each $k \geqslant 1$, choose $(z_k, \overrightarrow{u_k})$ in $T^1 \mathbb{D}$ such that $(u_k(-\infty), u_k(+\infty)) = F(U_k)$. It follows from Property 2.1 that the element $\pi^1((z_k, \overrightarrow{u_k}))$ is periodic. Furthermore, since F is continuous,

$$\lim_{k \to +\infty} (u_k(-\infty), u_k(+\infty)) = (u(-\infty), u(+\infty)).$$

Thus there exists a sequence $(s_k)_{k \geqslant 1}$ in $\mathbb{R}$ such that $(g_{s_k}(\pi^1((z_k, \overrightarrow{u_k}))))_{k \geqslant 1}$ converges to $\pi^1((z, \overrightarrow{u}))$.

2.2 An alternate proof of Theorem III.4.2

In order to prove the existence of geodesic trajectories which are dense in $\Omega_g(T^1 S)$, by Proposition 1.6 it is sufficient to prove the existence of T-orbits that are dense on Σ. Let us characterize sequences $S \in \Sigma$ such that $\overline{T^{\mathbb{Z}} S} = \Sigma$.

Let $V = (V_i)_{i \in I}$, with $I \subset \mathbb{Z}$, be a finite or infinite sequence with terms in $\mathcal{A}$. By definition, a *block of* V is a finite sequence B composed of consecutive terms of V. Equivalently B is of the form $B = (V_{n+i})_{1 \leqslant i \leqslant k}$, with $n + i \in I$ for each $1 \leqslant i \leqslant k$.

Property 2.2. *Let $S \in \Sigma$. Then the following are equivalent:*

(i) *all reduced words $a_1 \cdots a_n$ with $a_i \in \mathcal{A}$ and $a_{i+1} \neq a_i^{-1}$ are a block of S;*
(ii) $\overline{T^{\mathbb{Z}}(S)} = \Sigma$.

Proof.
(i) $\Rightarrow$ (ii). Let $S' = (S'_i)_{i \in \mathbb{Z}} \in \Sigma$. Consider the reduced word

$$S'_{-n} S'_{-n+1} \cdots S'_0 S'_1 \cdots S'_n.$$

For each n, by hypothesis there exists $k_n \in \mathbb{Z}$ such that $T^{k_n}(S)$ is a sequence U_n satisfying $U_{n,i} = S'_i$ for all $-n \leqslant i \leqslant n$. It follows that $\Delta(S', U_n) \leqslant \sqrt{2}/n$, which further implies that $\lim_{n \to +\infty} T^{k_n}(S) = S'$.
(ii) $\Rightarrow$ (i). Let $m = a_1 \cdots a_n$ be a reduced word and c be in $\mathcal{A} - \{a_n^{-1}, a_1^{-1}\}$. Consider the doubly-infinite periodic sequence S' having period $n+1$, defined by

$$S'_1 = a_1, \quad S'_2 = a_2, \quad S'_n = a_n, \quad S'_{n+1} = c.$$

This sequence belongs to Σ. Since $\overline{T^{\mathbb{Z}}(S)} = \Sigma$, there exists $(k_p)_{p \geqslant 1}$ in $\mathbb{N}$ such that $\lim_{p \to +\infty} T^{k_p}(S) = S'$. Define $T^{k_p}(S) = U_p$. For large enough p, $\Delta(S', U_p) \leqslant 2/(n+1)$. Thus $U_{p,i} = S'_i$ for all $1 \leqslant i \leqslant n+1$. This shows that the finite sequence $a_1, a_2, \ldots, a_n$ is a block of S. $\qquad\square$

A proof of Theorem III.4.2 using symbolic dynamics

To prove Theorem III.4.2, it remains to construct a sequence satisfying part (ii) of the above property. For each n in $\mathbb{N}^*$, let E_n denote the set of reduced words $m_n = a_1 \cdots a_n$ of length n, with $a_i \in \mathcal{A}$ and $a_i \neq a_{i+1}^{-1}$. Let ℓ_n be the number of elements of this set. Choose an enumeration $(m_{n,i})_{1 \leqslant i \leqslant \ell_n}$ of the elements of E_n. For all $1 \leqslant i < \ell_n$, choose a letter a_i in $\mathcal{A}$ which is not the inverse of either the last letter of the word $m_{n,i}$ or the first letter of the word $m_{n,i+1}$. Chosen this way, the word $m_{n,1} a_1 m_{n,2} a_2 \cdots a_{\ell_n-1} m_{n,\ell_n}$ is a reduced word. Let $B_n = (B_{n,i})_{1 \leqslant i \leqslant p_n}$ denote the finite sequence of letters forming this word. Choose $d_0 \neq B_{1,1}^{-1}$ in $\mathcal{A}$. For $n \geqslant 1$, choose d_n in $\mathcal{A} - \{B_{n,p_n}^{-1}, B_{n+1,1}^{-1}\}$. Finally, we are ready to present the doubly-infinite sequence S defined by $S_i = d_0$ for all $i \leqslant 0$ and whose sequence S^+ is constructed from blocks

$(B_n)_{n \geqslant 1}$ and the sequence $(d_n)_{n \geqslant 1}$ in the following way:

$$S^+ = \underbrace{B_{1,1}, B_{1,2}, \ldots, B_{1,p_1}}_{B_1}, d_1, \underbrace{B_{2,1}, \ldots, B_{2,p_2}}_{B_2}, d_2, \ldots,$$

$$\underbrace{B_{n,1}, \ldots, B_{n,p_n}}_{B_n}, d_n, \underbrace{B_{n+1,1}, \ldots,}_{B_{n+1}}, \ldots$$

The sequence S belongs to Σ. Yet by construction, the sequence S^+ contains all finite reduced words. Hence, by Property 2.2, one has $\overline{T^{\mathbb{Z}}(S)} = \Sigma$.

3 Applications to the general case

In Chap. III, we pointed out the existence of geodesic trajectories that are either periodic or dense in the non-wandering set associated with the quotient of $T^1\mathbb{D}$ by some non-elementary Fuchsian group.

Having settled that question, we now focus on the complementary question: are there any trajectories in the non-wandering set which satisfy neither of these two properties? To answer this question, we will use the fact that a non-elementary group contains some Schottky groups $S(g_1, g_2)$ generated by two hyperbolic isometries (Corollary II.1.3).

Proposition 3.1. *Let $S(g_1, g_2)$ be a Schottky group generated by two hyperbolic isometries. The non-wandering set of the geodesic flow on $S(g_1, g_2) \backslash T^1\mathbb{D}$ contains geodesic trajectories which are neither dense nor periodic.*

Exercise 3.2. Prove Proposition 3.1.
(Hint: construct an example of a non-periodic doubly-infinite sequence belonging to Σ which does not use all of the letters of the alphabet $\mathcal{A}$, and apply Proposition 1.6.)

Let Γ be a non-elementary Fuchsian group and $S(g_1, g_2)$ be a Schottky group included in Γ. Set $T^1 S_0 = S(g_1, g_2) \backslash T^1 \mathbb{D}$ and $T^1 S = \Gamma \backslash T^1 \mathbb{D}$. Consider the projection

$$P : T^1 S_0 \longrightarrow T^1 S.$$

Property 3.3. *The projection P satisfies the following properties:*

(i) *For all $t \in \mathbb{R}$ and $\pi^1((z, \overrightarrow{u})) \in T^1 S$, one has $P(g_t(\pi^1((z, \overrightarrow{u})))) = g_t(P(\pi^1((z, \overrightarrow{u}))))$.*
(ii) *If $P(\Omega_g(T^1 S_0)) = \Omega_g(T^1 S)$, then $L(\Gamma) = L(S(g_1, g_2))$.*

Exercise 3.4. Prove Property 3.3.

Lemma 3.5. *If a geodesic trajectory in $\Omega_g(T^1 S_0)$ is not periodic, then its image by P is not periodic.*

Proof. Let $g_{\mathbb{R}}(\pi^1((z, \overrightarrow{u})))$ be a non-periodic trajectory in $\Omega_g(T^1S_0)$. Suppose that its image by P is periodic. There exist a hyperbolic isometry γ in $\Gamma - S(g_1, g_2)$ and a real number $T \neq 0$ such that $\gamma((z, \overrightarrow{u})) = \widetilde{g}_T((z, \overrightarrow{u}))$. On the other hand, since $g_{\mathbb{R}}(\pi^1((z, \overrightarrow{u})))$ is included in a compact set, there exist an unbounded sequence $(t_n)_{n \geqslant 1}$ and a sequence $(\gamma_n)_{n \geqslant 1}$ in $S(g_1, g_2)$ for which the sequence $\gamma_n \widetilde{g}_{t_n}((z, \overrightarrow{u}))$ converges to an element in $T^1\mathbb{D}$. Using γ, one obtains a unbounded sequence of integers $(k_n)_{n \geqslant 1}$, and a bounded real sequence $(s_n)_{n \geqslant 1}$ such that $(\gamma_n \gamma^{k_n} \widetilde{g}_{s_n}((z, \overrightarrow{u})))_{n \geqslant 1}$ converges in $T^1\mathbb{D}$. The group Γ being discrete, the set of $\gamma_n \gamma^{k_n}$ is finite, this implies that γ^k is in $S(g_1, g_2)$ for some $k \neq 0$, which contradicts the fact that $g_{\mathbb{R}}(\pi^1((z, \overrightarrow{u})))$ is not periodic. $\quad\square$

Corollary 3.6. *Let Γ be a non-elementary Fuchsian group. The non-wandering set of the geodesic flow on $\Gamma \backslash T^1\mathbb{D}$ contains geodesic trajectories which are neither dense nor periodic.*

Proof. We choose $S(g_1, g_2)$ sufficiently "small" so that: $L(\Gamma) \neq L(S(g_1, g_2))$. It follows from Property 3.3, that the set $P(\Omega_g(T^1S_0))$ is a proper compact subset of $\Omega_g(T^1S)$ which is invariant with respect to the geodesic flow. Take the image by P of a geodesic trajectory on $\Omega_g(T^1S_0)$, which is neither periodic nor dense (Proposition 3.1). This image is a geodesic trajectory which not dense in $\Omega_g(T^1S)$ and not periodic (Lemma 3.5). $\quad\square$

We focus now on the existence of minimal compact sets which are invariant with respect to the geodesic flow on $\Omega_g(T^1S)$.

Recall that a subset F of a topological space is minimal relative to a group H of homeomorphisms if it is closed, non-empty, H-invariant and minimal in the sense of inclusion for these properties (Appendix A). Such a set is necessarily the closure of an orbit of H.

A decreasing sequence of compact sets which are invariant with respect to $g_{\mathbb{R}}$ contains a smaller element. Hence all compact subsets of $\Omega_g(T^1S)$ which are invariant with respect to the geodesic flow contain a minimal subset. This implies that, if the set $\Omega_g(T^1S)$ is compact, then it contains minimal sets. This general argument does not guarantee the existence of non-periodic minimal sets. We will prove that such sets do exist in $\Omega_g(T^1S)$. First we consider the case where $\Gamma = S(g_1, g_2)$ and use the doubly-infinite sequences introduced in Sect. 2.

Property 3.7. *Let $S = (S_i)_{i \in \mathbb{Z}}$ be an element of Σ. If for any n in $\mathbb{N}^*$, there exists $N(n) > 0$ such that for any integer j, the sequence $S_{-n}, S_{-n+1}, \ldots,$ $S_0, S_1, \ldots, S_n$ is a block of the sequence $S_{j+1}, \ldots, S_{j+N(n)}$, then $\overline{T^{\mathbb{Z}}(S)}$ is a minimal set for T.*

Proof. We are going to show that if S' belongs to $\overline{T^{\mathbb{Z}}(S)}$, then S belongs to $\overline{T^{\mathbb{Z}}(S')}$. This will show that $\overline{T^{\mathbb{Z}}(S)}$ is minimal.

Let $(p_k)_{k \geqslant 1}$ be a sequence in $\mathbb{Z}$ for which $\lim_{k \to +\infty} T^{p_k}(S) = S'$. One has $T^{p_k}(S) = (S_{p_k+i})_{i \in \mathbb{Z}}$. Fix n in $\mathbb{N}^*$. Let $N(n)$ be the integer associated to n.

Since $\lim_{k\to+\infty} T^{p_k}(S) = S'$, there exists $k_n > 0$ such that $S_{p_{k_n}+i} = S'_i$ for all $1 \leqslant i \leqslant N(n)$. It follows that there exists $0 \leqslant j_n \leqslant N(n) - 2n - 1$ such that $S'_{j_n+i} = S_{-n+i-1}$ for all $1 \leqslant i \leqslant 2n + 1$. Thus $\Delta(T^{j_n+n+1}(S'), S) \leqslant \sqrt{2}/n$. This shows that $\lim_{n\to+\infty} T^{j_n+n+1}(S') = S$. $\qquad\square$

It remains to construct a sequence in Σ satisfying Property 3.7. For this purpose, consider the sequence of words $(m_n)_{n\geqslant 1}$ defined recursively by

$$m_1 = g_1 \quad \text{and} \quad m_{n+1} = m_n g_2 m_n m_n g_2 m_n,$$

where g_1, g_2 are the generators of $S(g_1, g_2)$. Notice that in each step, each word m_n begins and ends with the letter g_1 and therefore each is a reduced word. Let ℓ_n denote the length of the word m_n. The ℓ_n initial letters of m_{n+1} coincide with those of m_n. The first letter of m_n is also different from the inverse of its last letter. Thus there exists a (unique) doubly-infinite sequence $S = (S_i)_{i\in\mathbb{Z}}$, with $S_i \in \mathcal{A}$ satisfying the conditions

$$S_{-\ell_n+1}S_{-\ell_n+2}\cdots S_0 S_1 \cdots S_{\ell_n} = m_n m_n.$$

Exercise 3.8. Prove that S belongs to Σ and is not periodic.

Lemma 3.9. *Let $n \geqslant 1$. For any $p \geqslant n+1$, all blocks of length $6\ell_n + 3$ of the sequence $S_{-\ell_p+1}, \ldots, S_0, S_1, \ldots, S_{\ell_p}$ contain the block $S_{-\ell_n}, S_{-\ell_n+1}, \ldots, S_0, S_1, \ldots, S_{\ell_n}$.*

Proof. Fix $n \geqslant 1$. Define $A_p = m_p m_p$ and proceed by induction on $p \geqslant n+1$.

One has $A_{n+1} = m_n g_2 A_n g_2 A_n g_2 A_n g_2 m_n$. The length of the sequences associated to m_n and A_n being respectively ℓ_n and $2\ell_n$, all blocks of length $6\ell_n + 3$ of the sequence associated to A_{n+1} contain the sequence associated to A_n, which proves the property for $p = n+1$.

Take some $p \geqslant n+1$. Assume now that the property is true for this p and let us show that it is true up to $p+1$.

Consider $A_{p+1} = m_p g_2 A_p g_2 A_p g_2 A_p g_2 m_p$. Choose a block B of length $6\ell_n + 3$ of the sequence associated to A_{p+1}. If B is a block of A_p or of m_p, the induction hypothesis applies. Otherwise B is a block of the sequence associated to one of the following words w_i for $i = 1, 2, 3$ defined as follow:

$$w_1 = m_p g_2 A_p, \qquad w_2 = A_p g_2 A_p, \qquad w_3 = A_p g_2 m_p.$$

Such a block must contain the letter g_2 written in one of the words above. Let $\overline{g}_2$ denote this copy of g_2. By construction, any w_i contains

$$M_n = A_n g_2 m_n \overline{g_2} m_n g_2 A_n.$$

Since $m_p = m_{p-1} g_2 A_{p-1} g_2 m_{p-1}$ and $A_p = m_p m_p$, the sequence associated to the word M_n is a block of length $6\ell_n + 3$ of the sequence associated to the words w_1, w_2, w_3. Since block B has length $6\ell_n + 3$ and B contains $\overline{g}_2$, B contains the sequence associated to A_n. $\qquad\square$

Corollary 3.10. *The closed set* $\overline{T^{\mathbb{Z}}(S)}$ *is minimal and non-periodic.*

Proof. We show that Property 3.7(i) is satisfied. Let us fix $n \in \mathbb{N}^*$. For all $j \in \mathbb{N}^*$, there exists $k \geqslant n+1$ such that the finite sequence $B = S_{j+1}, \ldots, S_{j+6\ell_n+3}$ is a block of $S_{-\ell_k+1}, \ldots, S_0, S_1, \ldots, S_{\ell_k}$. Lemma 3.9 implies that the block B contains $S_{-\ell_n}, \ldots, S_0, \ldots, S_{\ell_n}$, thus in particular it contains $S_{-n}, \ldots, S_0, \ldots, S_n$. Furthermore, by Exercise 3.8, S is not periodic. $\qquad\square$

Let us return to the geodesic flow on $\Omega_g(S(g_1, g_2)\backslash T^1\mathbb{D})$. Let $(z, \overrightarrow{u}) \in T^1\mathbb{D}$ be such that $(u(-\infty), u(+\infty)) = F(S)$, where S is given by Corollary 3.10. Notice that, since S is not periodic, neither is $\pi^1((z, \overrightarrow{u}))$.

Exercise 3.11. Let Γ be a Fuchsian group. Prove that a compact trajectory for the geodesic flow on $\Omega_g(\Gamma\backslash T^1\mathbb{D})$ is periodic.

Since $\Omega_g(S(g_1, g_2)\backslash T^1\mathbb{D})$ is compact and $\pi^1((z, \overrightarrow{u}))$ is not periodic, it follows from Exercise 3.11 that $g_{\mathbb{R}}(\pi^1((z, \overrightarrow{u})))$ is not closed.

Theorem 3.12. *The set* $\overline{g_{\mathbb{R}}(\pi^1((z, \overrightarrow{u})))} \subset \Omega_g(S(g_1, g_2)\backslash T^1\mathbb{D})$ *is a compact minimal set relative to* $g_{\mathbb{R}}$, *which is not periodic.*

Proof. Let us show that $\overline{g_{\mathbb{R}}(\pi^1((z, \overrightarrow{u})u))}$ is minimal relative to $g_{\mathbb{R}}$. Take $\pi^1((z', \overrightarrow{u}')) \in \overline{g_{\mathbb{R}}(\pi^1((z, \overrightarrow{u})))}$. By Lemma 1.5, after replacing $(z', \overrightarrow{u}')$ with an element of $\Gamma(z', \overrightarrow{u}')$, one can assume that there exists a sequence S' in Σ such that $(u'(-\infty), u'(+\infty)) = F(S')$. By Proposition 1.6, one has $S' \in \overline{T^{\mathbb{Z}}(S)}$. The set $\overline{T^{\mathbb{Z}}(S)}$ is minimal and S belongs to $\overline{T^{\mathbb{Z}}(S')}$, therefore $\pi^1((z, \overrightarrow{u}))$ belongs to $\overline{g_{\mathbb{R}}(\pi^1((z', \overrightarrow{u}')))}$. This shows that $\overline{g_{\mathbb{R}}(\pi^1((z, \overrightarrow{u})))}$ is minimal. $\qquad\square$

Consider now a non-elementary Fuchsian group Γ.

Corollary 3.13. *The set* $\Omega_g(\Gamma\backslash T^1\mathbb{D})$ *contains compact minimal sets which are non-periodic for the geodesic flow.*

Proof. Choose a Schottky subgroup $S(g_1, g_2)$ of Γ. Consider a compact minimal non-periodic set $K \subset \Omega_g(S(g_1, g_2)\backslash T^1\mathbb{D})$ given by Theorem 3.12. The projection P sends K to a compact subset K' of $\Omega_g(\Gamma\backslash T^1\mathbb{D})$ which is invariant with respect to the geodesic flow (Property 3.3). It is minimal. To see this, suppose that K' properly contains a compact non-empty K_0 which is invariant with respect to $g_{\mathbb{R}}$. The set $P^{-1}(K_0) \cap K$ would therefore be a compact non-empty $g_{\mathbb{R}}$-invariant proper subset of K. This is impossible since K is minimal. Moreover, K' is not minimal, according to Lemma 3.5. $\qquad\square$

4 Comments

This approach to geodesic flow by way of symbolic dynamics was originally developed for the modular surface [2, 3, 20, 57]. It extends to quotients of the Poincaré half-plane by geometrically finite Fuchsian groups [55].

This point of view can be generalized to quotients of pinched Hadamard manifolds by some cocompact Kleinian groups (method of Markov partitions [58]) and by some Schottky groups [21, 44].

The coding relates the ergodic theory of geodesic flow to that of sub-shifts of finite type and Ruelle-Perron-Frobenius operators. For example, it allows one to recover the Gauss measure $dx/(1 + x)$ on $[0, 1]$ which is invariant with respect to the Gauss function $t(x) = 1/x - [1/x]$, from the Liouville measure on $\mathrm{PSL}(2, \mathbb{Z})\backslash T^1 \mathbb{H}$ which is invariant with respect to the geodesic flow [20, 57].

This symbolic method is especially effective for calculating the entropy of the geodesic flow or enumerating the closed geodesics of a manifold [44, 21]. The book by T. Bedford, M. Keane & C. Series [8] is a good reference for this approach and its applications. As we saw in Sect. 3, this coding also allows us to construct some minimal sets for the geodesic flow. This construction is due to Morse and is described in the book by W. Gottschalk and G. Hedlund [34].

The sets that we have constructed are compact. In [22], we present a construction of non-trivial non-compact minimal sets, when the surface has cusps.

V

Topological dynamics of the horocycle flow

In this chapter, we analyze the topology of the trajectories of another classical example of flow on the quotient of $T^1\mathbb{H}$ by a Fuchsian group: the *horocycle flow*. Our method is based on a correspondence between the set of horocycles of $\mathbb{H}$ and the space of non-zero vectors in $\mathbb{R}^2$ modulo $\{\pm\,\mathrm{Id}\}$. This vectorial point of view allows one to relate the topological dynamics of the linear action on $\mathbb{R}^2$ of a discrete subgroup Γ of $\mathrm{SL}(2,\mathbb{R})$ to that of the horocycle flow on the quotient of $T^1\mathbb{H}$ by the Fuchsian group corresponding to Γ. In the geometrically finite case, we show that the horocycle flow is less topologically turbulent than the geodesic flow (Sect. 4).

Throughout this chapter, we use the definitions and notations associated with the dynamics of a flow as originally introduced in Appendix A.

1 Preliminaries

1.1 The horocycle flow on $T^1\mathbb{H}$

To each element $(z, \overrightarrow{u}) \in T^1\mathbb{H}$, we associate the horocycle H passing through z centered at $u(+\infty)$. Let $\beta : \mathbb{R} \to H$, be the arclength parametrization of H such that: $\beta(0) = z$, and the pair of vectors $(d\beta/ds(0), \overrightarrow{u})$ forms a positive basis for $T_z\mathbb{H}$ (Fig. V.1). The image of β is called the *oriented horocycle associated to* $(z, \overrightarrow{u})$.

Exercise 1.1. Prove that if $z = a+ib$ and $u(+\infty) = \infty$, then $\beta(s) = a+sb+ib$ for all $s \in \mathbb{R}$ (Fig. V.2).

For fixed t, we introduce the function $\widetilde{h}_t : T^1\mathbb{H} \to T^1\mathbb{H}$ defined by

$$\widetilde{h}_t((z, \overrightarrow{u})) = \big(\beta(t), \overrightarrow{v}(t)\big),$$

where $\overrightarrow{v}(t)$ is the unit vector in $T^1_{\beta(t)}\mathbb{H}$ for which the pair $(d\beta/ds(t), \overrightarrow{v}(t))$ is a positive orthonormal basis (Fig. V.3).

F. Dal'Bo, *Geodesic and Horocyclic Trajectories*, Universitext,
DOI 10.1007/978-0-85729-073-1_5, © Springer-Verlag London Limited 2011

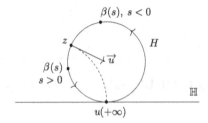

Fig. V.1.

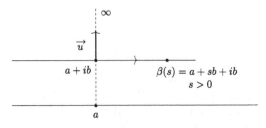

Fig. V.2.

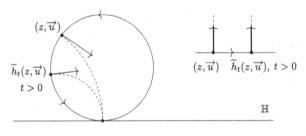

Fig. V.3.

Exercise 1.2. Prove that for all positive isometries $g \in G$ and for all real numbers t, the following equality holds:

$$g \circ \tilde{h}_t = \tilde{h}_t \circ g.$$

Exercise 1.3. Prove that $\tilde{h}_t$ is a homeomorphism of $T^1\mathbb{H}$ equipped with the metric D.
(Hint: using Property I.2.3, Exercises III.1.1 and 1.2, prove that if a sequence $((z_n, \overrightarrow{u_n}))_{n \geqslant 1}$ in $T^1\mathbb{H}$ converges to $(z, \overrightarrow{u})$, then $(\tilde{h}_t((z_n, \overrightarrow{u_n})))_{n \geqslant 1}$ converges to $\tilde{h}_t((z, \overrightarrow{u}))$.)

For all t and t' in $\mathbb{R}$, one has

$$\tilde{h}_{t+t'} = \tilde{h}_t \circ \tilde{h}_{t'}.$$

To see this, take $(i, \vec{u})$ in $T^1\mathbb{H}$ such that $u(+\infty) = \infty$. By Exercises 1.1 and 1.2 one has, for all positive isometries g,

$$\widetilde{h}_{t+t'}(g((i, \vec{u}))) = g((z', \vec{u}')) \quad \text{with } z' = i+t+t' \text{ and } u'(+\infty) = \infty.$$

Yet $(z', \vec{u}') = \widetilde{h}_t(\widetilde{h}_{t'}((i, \vec{u})))$. Therefore,

$$\widetilde{h}_{t+t'}(g((i, \vec{u}))) = \widetilde{h}_t(\widetilde{h}_{t'}(g((i, \vec{u})))).$$

Since the group of positive isometries of $\mathbb{H}$ acts transitively on $T^1\mathbb{H}$ (Property I.2.3), the above statement is satisfied by all elements of $T^1\mathbb{H}$.

It follows that the map from $(\mathbb{R}, +)$ into the group of homeomorphisms of $T^1\mathbb{H}$ which sends t to $\widetilde{h}_t$ is a group homomorphism.

Exercise 1.4. Prove that for all $t, t' \in \mathbb{R}$ and $(z, \vec{u})$ in $T^1\mathbb{H}$, one has

$$D(\widetilde{h}_t((z, \vec{u})), \widetilde{h}_{t'}((z, \vec{u}))) = 4\ln\left(\frac{1}{2}\left(|t'-t| + \sqrt{|t'-t|^2+4}\right)\right).$$

Exercises 1.3 and 1.4 imply that the map $\widetilde{h} : \mathbb{R} \times T^1\mathbb{H} \to T^1\mathbb{H}$ defined by

$$\widetilde{h}(t, (z, \vec{u})) = \widetilde{h}_t((z, \vec{u}))$$

is continuous. This map defines a flow (Appendix A) on $T^1\mathbb{H}$, that we call the *horocycle flow*.

As in the case of the geodesic flow (see Exercise III.1.4), the dynamics of this flow on $T^1\mathbb{H}$ are fairly straightforward.

Exercise 1.5. Prove the following properties:

(i) the set $\Omega_{\widetilde{h}}(T^1\mathbb{H})$ is empty;
(ii) for all $(z, \vec{u})$ in $T^1\mathbb{H}$, the map from $\mathbb{R}$ into $T^1\mathbb{H}$ which sends t to $\widetilde{h}_t((z, \vec{u}))$ is an embedding.

1.2 A vectorial point of view on the space of trajectories of $\widetilde{h}_\mathbb{R}$

Let E be the quotient of $\mathbb{R}^2 - \{(0,0)\}$ over $\{\pm\mathrm{Id}\}$. Consider the map (Fig. V.4)

$$\mathrm{vect} : T^1\mathbb{H} \longrightarrow E$$

defined by

$$(z, \vec{u}) \mapsto \mathrm{vect}((z, \vec{u}))$$
$$= \begin{cases} \pm e^{B_{u(+\infty)}(i,z)/2}/\sqrt{1+u^2(+\infty)}\binom{u(+\infty)}{1} & \text{if } u(+\infty) \neq \infty, \\ \pm e^{B_{u(+\infty)}(i,z)/2}\binom{1}{0} & \text{if } u(+\infty) = \infty. \end{cases}$$

Note that if $u(+\infty) = \infty$ and $z = i$, then $\mathrm{vect}((z, \vec{u})) = \pm\binom{1}{0}$. Clearly, we have

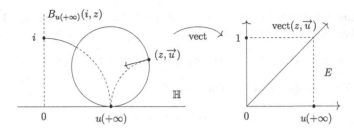

Fig. V.4.

Property 1.6. *The map* vect *is surjective, constant on the trajectories of* $\widetilde{h}_{\mathbb{R}}$.

The map vect induces an action on E of the group G of positive isometries of $\mathbb{H}$. The following proposition clarifies the nature of this action. Let us introduce some notation.

For all $g \in G$, set $g(z) = (az + b)/(cz + d)$, with $ad - bc = 1$, and let M_g denote the linear transformation acting on E defined by

$$\forall \pm \begin{pmatrix} x \\ y \end{pmatrix} \in E = \{\pm \mathrm{Id}\}\backslash(\mathbb{R}^*)^2, \quad M_g\left(\pm\begin{pmatrix} x \\ y \end{pmatrix}\right) = \pm \begin{pmatrix} a & b \\ c & d \end{pmatrix}\begin{pmatrix} x \\ y \end{pmatrix}.$$

Proposition 1.7. *For all g in G and $(z, \overrightarrow{u})$ in $T^1\mathbb{H}$, one has*

$$\mathrm{vect}(g((z, \overrightarrow{u}))) = M_g(\mathrm{vect}((z, \overrightarrow{u}))).$$

Proof. Consider $(i, \overrightarrow{u_1})$ in $T^1\mathbb{H}$ defined by $u_1(+\infty) = \infty$. Recall that $\mathrm{vect}((i, \overrightarrow{u_1})) = \pm e_1$, where $e_1 = \begin{pmatrix} 1 \\ 0 \end{pmatrix}$. We first show that for all g in G, $\mathrm{vect}(g((i, \overrightarrow{u_1}))) = M_g(\pm e_1)$. Let us decompose $M_g = \pm \begin{pmatrix} a & b \\ c & d \end{pmatrix}$ as $M_g = \pm KAN$, where

$$K = \begin{pmatrix} \cos\theta & -\sin\theta \\ \sin\theta & \cos\theta \end{pmatrix}, \quad A = \begin{pmatrix} \lambda & 0 \\ 0 & \lambda^{-1} \end{pmatrix}, \quad \lambda > 0, \quad N = \begin{pmatrix} 1 & t \\ 0 & 1 \end{pmatrix}$$

(Proposition I.2.4). Let k and a be the Möbius transformations associated with K and A. We have

$$B_\infty(i, g^{-1}(i)) = B_\infty(i, a^{-1}(i)) \quad \text{and} \quad B_\infty(i, a^{-1}(i)) = B_\infty(i, \lambda^{-2}i).$$

Hence $B_\infty(i, g^{-1}(i)) = \ln \lambda^{-2}$. Furthermore, $\|M(e_1)\| = \lambda$, which implies that $\|\mathrm{vect}(g((i, \overrightarrow{u_1})))\| = \|M_g(\pm e_1)\|$. Additionally, $g(u_1(+\infty)) = k(\infty)$ and $M(e_1)$ is colinear with $K(e_1)$. These two facts prove the desired equality in the case where $(z, \overrightarrow{u}) = (i, \overrightarrow{u_1})$.

We now show that, for all $(z, \overrightarrow{u})$ in $T^1\mathbb{H}$,

$$\mathrm{vect}(g((z, \overrightarrow{u}))) = M_g\,\mathrm{vect}((z, \overrightarrow{u})).$$

Since the action of G on $T^1\mathbb{H}$ is simply transitive (Property I.2.3), there exists g' in G such that $g'((i, \overrightarrow{u_1})) = (z, \overrightarrow{u})$. Therefore, $\mathrm{vect}(g((z, \overrightarrow{u}))) = \mathrm{vect}(gg'((i, \overrightarrow{u_1})))$, and hence

$$\mathrm{vect}(g((z, \overrightarrow{u}))) = M_{gg'}(\mathrm{vect}((i, \overrightarrow{u_1}))).$$

At this point, it is sufficient to observe that

$$M_{gg'}(\mathrm{vect}((i, \overrightarrow{u_1}))) = M_g(M_{g'}(\mathrm{vect}((i, \overrightarrow{u_1}))))$$

and $M_{g'}(\mathrm{vect}((i, \overrightarrow{u_1}))) = \mathrm{vect}(g'((i, \overrightarrow{u_1})))$. $\qquad\square$

Exercise 1.8. Prove that the map vect is continuous.
(Hint: Use Property I.2.2 and Proposition 1.7.)

The following property provides a further characterization of the horocyclic and parabolic points of the limit set of a Fuchsian group (Sect. I.3.2).

Property 1.9. *Let $(z, \overrightarrow{u})$ be in $T^1\mathbb{H}$ and $(g_n)_{n \geqslant 1}$ be a sequence in G.*

(i) *The sequence $(B_{u(+\infty)}(i, g_n^{-1}(i)))_{n \geqslant 1}$ tends to $+\infty$ if and only if $(\|M_{g_n}(\mathrm{vect}((z, \overrightarrow{u})))\|)_{n \geqslant 1}$ converges to 0.*
(ii) *The point $u(+\infty)$ is fixed by a parabolic isometry $g \in G - \{\mathrm{Id}\}$ if and only if $M_g(\mathrm{vect}((z, \overrightarrow{u}))) = \mathrm{vect}((z, \overrightarrow{u}))$.*

Exercise 1.10. Prove Property 1.9.

2 The horocycle flow on a quotient

Let us consider a Fuchsian group Γ. We will retain the notation introduced in Sect. III.1.2. The commutativity of $\widetilde{h}_\mathbb{R}$ and G proved in Exercise 1.2, allows one to define the *horocycle flow* $h_\mathbb{R}$ on the quotient $T^1S = \Gamma \backslash T^1\mathbb{H}$ (Fig. V.5): for all $(z, \overrightarrow{u})$ in $T^1\mathbb{H}$, we set

$$h_t(\pi^1((z, \overrightarrow{u}))) = \pi^1(\widetilde{h}_t((z, \overrightarrow{u}))).$$

By definition of the topology of T^1S, a sequence $(\pi^1((z_n, \overrightarrow{u_n})))_{n \geqslant 1}$ converges to $\pi^1((z, \overrightarrow{u}))$ if and only if there exists a sequence $(\gamma_n)_{n \geqslant 1}$ in Γ such that $\lim_{n \to +\infty} \gamma_n(z_n) = z$ and $\lim_{n \to +\infty} \gamma_n(u_n(+\infty)) = u(+\infty)$.

Thus $(h_{t_n}(\pi^1((z, \overrightarrow{u}))))_{n \geqslant 1}$ converges to $\pi^1((z', \overrightarrow{u}'))$ if and only if there exists $(\gamma_n)_{n \geqslant 1}$ in Γ such that

$$\lim_{n \to +\infty} \gamma_n z_n = z' \quad \text{and} \quad \lim_{n \to +\infty} \gamma_n(u(+\infty)) = u'(+\infty),$$

where z_n is the projection on $\mathbb{H}$ of $\widetilde{h}_{t_n}((z, \overrightarrow{u}))$.

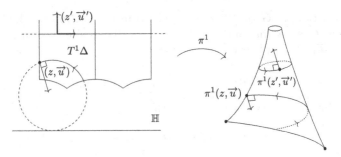

Fig. V.5. $\Gamma = \mathrm{PSL}(2, \mathbb{Z})$

2.1 A vectorial point of view on $h_{\mathbb{R}}$

Let M_Γ denote the subgroup of $\{\pm \mathrm{Id}\} \backslash \mathrm{SL}(2, \mathbb{R})$ consisting of all M_γ with γ in Γ. This group of linear transformations is isomorphic to Γ. The following proposition relates its dynamics on E to those of $h_{\mathbb{R}}$ on $T^1 S$.

Proposition 2.1. *Let* $(z, \overrightarrow{u})$ *and* $(z', \overrightarrow{u}')$ *be in* $T^1 \mathbb{H}$. *There exists a sequence* $(t_n)_{n \geqslant 1}$ *in* $\mathbb{R}$ *such that* $(h_{t_n}(\pi^1((z, \overrightarrow{u}))))_{n \geqslant 1}$ *converges to* $\pi^1((z', \overrightarrow{u}'))$ *if and only if there exists a sequence* $(M_{\gamma_n})_{n \geqslant 1}$ *in* M_Γ *such that* $(M_{\gamma_n}(\mathrm{vect}((z, \overrightarrow{u}))))_{n \geqslant 1}$ *converges to* $\mathrm{vect}((z', \overrightarrow{u}'))$.

Proof. Assume that there exists $(\gamma_n)_{n \geqslant 1}$ in Γ for which $(\gamma_n \widetilde{h}_{t_n}((z, \overrightarrow{u})))_{n \geqslant 1}$ converges to $(z', \overrightarrow{u}')$. Its image by the map vect converges to $\mathrm{vect}((z', \overrightarrow{u}'))$, since vect is continuous. We have

$$\mathrm{vect}(\gamma_n \widetilde{h}_{t_n}((z, \overrightarrow{u}))) = M_{\gamma_n}(\mathrm{vect}((z, \overrightarrow{u}))),$$

hence $(M_{\gamma_n}(\mathrm{vect}((z, \overrightarrow{u}))))_{n \geqslant 1}$ converges to $\mathrm{vect}((z', \overrightarrow{u}'))$.

Conversely, assume that $(M_{\gamma_n}(\mathrm{vect}((z, \overrightarrow{u}))))_{n \geqslant 1}$ converges to $\mathrm{vect}((z', \overrightarrow{u}'))$. By definition of the map vect, the sequences $(\gamma_n(u(+\infty)))_{n \geqslant 1}$ and $(B_{\gamma_n(u(+\infty))}(i, \gamma_n(z)))_{n \geqslant 1}$ converge to $u'(+\infty)$ and $B_{u'(+\infty)}(i, z')$ respectively. Consider the real number t_n such that $\widetilde{h}_{t_n}(\gamma_n((z, \overrightarrow{u})))$ is tangent to the geodesic passing through z' having $\gamma_n(u(+\infty))$ as an endpoint. Set $\widetilde{h}_{t_n}(\gamma_n((z, \overrightarrow{u}))) = (z_n, \overrightarrow{u_n})$. Since $u_n(+\infty) = \gamma_n(u(+\infty))$, the sequence $(u_n(+\infty))_{n \geqslant 1}$ converges to $u'(+\infty)$. Moreover $B_{\gamma_n(u(+\infty))}(i, \gamma_n(z)) = B_{u_n(+\infty)}(i, z_n)$, hence $(z_n)_{n \geqslant 1}$ converges to z'. One thus obtains

$$\lim_{n \to +\infty} h_{t_n}(\pi^1((z, \overrightarrow{u}))) = \pi^1((z', \overrightarrow{u}')). \qquad \square$$

Corollary 2.2. *Let* $(z, \overrightarrow{u})$ *be in* $T^1 \mathbb{H}$. *The trajectory* $h_{\mathbb{R}}(\pi^1((z, \overrightarrow{u})))$ *is closed in* $T^1 S$ *if and only if the orbit of the vector* $\mathrm{vect}((z, \overrightarrow{u}))$ *with respect to the group* M_Γ *is closed in* E.

Exercise 2.3. Prove Corollary 2.2.

2.2 Characterization of the non-wandering set

Let us introduce the subset of E defined by

$$E(\Gamma) = \{\text{vect}((z, \overrightarrow{u})) \mid (z, \overrightarrow{u}) \in T^1\mathbb{H}, u(+\infty) \in L(\Gamma)\}.$$

Exercise 2.4. Prove that the set $E(\Gamma)$ is closed in E, and invariant with respect to the action of the group M_Γ.

If $L(\Gamma) = \mathbb{H}(\infty)$, in particular if Γ is the modular group, then $E(\Gamma) = E$. Otherwise, $E(\Gamma)$ is a proper subset of E. For example, if Γ is a Schottky group, then $E(\Gamma)$ is homeomorphic to the product of $\mathbb{R}^*_+$ with a Cantor set.

Let $PE(\Gamma)$ denote the image of $E(\Gamma)$ by the projection from E to the projective line $\mathbb{RP}^1$.

Exercise 2.5. Prove that the set $PE(\Gamma) \subset \mathbb{RP}^1$ is closed and invariant with respect to the projective action of M_Γ on $\mathbb{RP}^1$.

Prove that this action, when conjugated by a homeomorphism, is identical to the action of Γ on $L(\Gamma)$.

(Hint: use Exercise I.1.14.)

It follows that, if Γ is not elementary, then every projective orbits of M_Γ on $PE(\Gamma)$ is dense; in other words, $PE(\Gamma)$ is a minimal set for this action.

Is this property also satisfied by the action of M_Γ on $E(\Gamma)$?

At least, in the case where Γ is the modular group, the answer is "No" since $E(\Gamma) = \{\pm\text{Id}\}\backslash\mathbb{R}^2$ and the quotient of $\mathbb{Z}^2 - \{(0,0)\}$ over $\{\pm\text{Id}\}$ is a closed subset of E, invariant under $\text{PSL}(2, \mathbb{Z})$.

In the next section, we will specify a necessary and sufficient condition on Γ to allow the answer to this question to be "Yes" (Proposition 4.3(ii)). In general, only the existence of dense orbits is guaranteed.

Proposition 2.6. *Let Γ be a non-elementary Fuchsian group. There exists $\pm w \in E(\Gamma)$ such that $\overline{M_\Gamma(w)} = E(\Gamma)$.*

To prove this proposition, we use the vector space $\mathbb{R}^2$ and consider the subgroup $\widehat{M}_\Gamma$ of $\text{SL}(2, \mathbb{R})$ which is the pre-image of M_Γ with respect to the projection of $\text{SL}(2, \mathbb{R})$ onto $\text{PSL}(2, \mathbb{R})$. This group acts on the set $\widehat{E}(\Gamma) \subset \mathbb{R}^2 - \{(0,0)\}$ which is the pre-image of $E(\Gamma)$ with respect to the projection of $\mathbb{R}^2 - \{(0,0)\}$ onto E.

Let us begin by proving the following lemma:

Lemma 2.7. *Let B_1 and B_2 be two open disks in $\mathbb{R}^2 - \{(0,0)\}$ which have non-empty intersection with $\widehat{E}(\Gamma)$. There exists M in $\widehat{M}_\Gamma$ such that $MB_1 \cap B_2 \neq \varnothing$.*

Proof. For $i = 1, 2$, we let C_i denote the positive open cone generated by B_i. The disk B_1 intersects $\widehat{E}(\Gamma)$, and the projective action of $\widehat{M}_\Gamma$ on $PE(\Gamma)$ is minimal (since Γ is non-elementary and according to Exercise 2.5), hence there exists an eigenvector u_1^+ in B_1 associated to $M_1 \in \widehat{M_\Gamma}$ which is associated with a hyperbolic isometry γ_1 in Γ. One can assume that $M_1(u_1^+) = \lambda_1 u_1^+$, where $\lambda_1 > 1$. Let γ be a hyperbolic isometry in Γ not having any fixed points in common with those of γ_1. Choose $M \in \widehat{M_\Gamma}$ which projects to the matrix associated with γ, and two eigenvectors u^+ and u^- of M such that $M(u^+) = \lambda u^+$, $M(u^+) = \lambda^{-1} u^-$, where $|\lambda| > 1$. After replacing M_1 by M_1^n and u^+, u^- by some vectors in $\mathbb{R}^* u^+, \mathbb{R}^* u^-$, one can assume that $M_1(u^+)$ and $M_1(u^-)$ belong to B_1. Let M_2 be an element of $\widehat{M_\Gamma}$ which projects to the matrix associated with a hyperbolic isometry in Γ whose eigen directions are distinct from those of M_1 and whose attractive eigenvectors belong to C_2. After replacing M_2 by $\pm M_2^n$, one can assume that $M_2(u^+)$ belongs to C_2. The segment $[M_2 M^n(u^-), M_2 M^n(u^+)]$ converges to the open ray Δ beginning at 0 in the direction of $M_2(u^+)$, as n tends to $+\infty$. This ray is also the limit of the images of the maps $M_2 M^n M_1^{-1}$ restricted to the segment $[M_1(u^-), M_1(u^+)]$. This limit is contained in B_1. Since $M_2(u^+)$ belongs to C_2, the ray Δ intersects the open set B_2 in a segment of non-zero length (i.e., not a point). Therefore, there exist w in B_1 and n such that $M_2 M^n M_1^{-1}(w)$ is in B_2.   □

Proof (of Proposition 2.6). We shall prove that there exists x in $\mathbb{R}^2 - \{(0,0)\}$ such that $\overline{\widehat{M_\Gamma}(x)} = \widehat{E}(\Gamma)$. The idea is the same as that used to prove Theorem III.4.2.

Let $(B_n)_{n \geqslant 1}$ be a sequence of open disks of $\mathbb{R}^2 - \{(0,0)\}$ each having non-empty intersection with $\widehat{E}(\Gamma)$, and such that any open subset of $\mathbb{R}^2 - \{(0,0)\}$ intersecting $\widehat{E}(\Gamma)$ contains a disk in this sequence. Fix such an open set O. By Lemma 2.7, there exists M_1 in $\widehat{M_\Gamma}$ such that $M_1(O) \cap B_1 \neq \varnothing$. Let K_1 be an open, pre-compact set intersecting $\widehat{E}(\Gamma)$, contained in O, such that $M_1(K_1)$ is in B_1. In the preceding argument, replacing O with K_1 and B_1 by B_2, one obtains $M_2 \in \widehat{M_\Gamma}$ and an open relatively compact set $K_2 \subset K_1$ intersecting $\widehat{E}(\Gamma)$ such that $M_2 K_2 \subset B_2$. In this way, one obtains two sequences $(M_n)_{n \geqslant 1}$ in $\widehat{M_\Gamma}$ and $(K_n)_{n \geqslant 1}$ of nested, pre-compact open sets intersecting $\widehat{E}(\Gamma)$ such that $M_n(K_n) \subset B_n$. Let x be in $\bigcap_{n=1}^{+\infty} K_n \cap \widehat{E}(\Gamma)$. For all $n \geqslant 1$, the point $M_n(x)$ is in B_n. Consider an element x' in $\widehat{E}(\Gamma)$ and a sequence $(D_n)_{n \geqslant 1}$ of disks centered at x' whose radius converges to 0. Each D_n contains a disk B_{i_n} thus $M_{i_n}(x)$ is in D_n. This shows that x' belongs to $\overline{\widehat{M_\Gamma}(x)}$ and thus that $\widehat{M_\Gamma}(x)$ is dense in $\widehat{E}(\Gamma)$.   □

Let us return to the horocycle flow and introduce the subset $\widetilde{F}(\Gamma)$ in $T^1 \mathbb{H}$, defined by

$$\widetilde{F}(\Gamma) = \text{vect}^{-1}(E(\Gamma)).$$

This set is also defined by

$$\widetilde{F}(\Gamma) = \{(z, \vec{u}) \in T^1 \mathbb{H} \mid u(+\infty) \in L(\Gamma)\}.$$

Exercise 2.8. Prove that $\widetilde{F}(\Gamma)$ is closed and invariant with respect to Γ and $\widetilde{h}_\mathbb{R}$.

Let $F(\Gamma)$ denote the image of $\widetilde{F}(\Gamma)$ by the projection π^1 to $T^1 S$. This set is closed and invariant with respect to the flow $h_\mathbb{R}$. Furthermore, it is a proper subset of $T^1 S$ if and only if $L(\Gamma) \neq \mathbb{H}(\infty)$.

Exercise 2.9. Prove that $F(\Gamma)$ is compact if and only if S is compact.

Notice that Propositions 2.1 and 2.6 together imply the following proposition.

Proposition 2.10. *There exists* $(z, \overrightarrow{u}) \in \widetilde{F}(\Gamma)$ *such that* $\overline{h_\mathbb{R}(\pi^1((z, \overrightarrow{u})))} = F(\Gamma)$.

Can $F(\Gamma)$ be characterized by some property of the horocycle flow? This question is answered by the following proposition.

Proposition 2.11. *The set* $F(\Gamma)$ *is the non-wandering set,* $\Omega_h(T^1 S)$, *of the horocycle flow on* $T^1 S$.

Proof. Assume that $\pi^1((z, \overrightarrow{u}))$ is non-wandering (see Appendix A). There is a sequence $(V_n)_{n \geqslant 1}$ of nested neighborhoods of $\pi^1((z, \overrightarrow{u}))$ and a sequence of positive real numbers $(t_n)_{n \geqslant 1} \subset \mathbb{R}^+$ such that $\lim_{n \to +\infty} t_n = +\infty$ and $h_{t_n}(V_n) \cap V_n \neq \varnothing$. Consider a sequence $((z_n, \overrightarrow{u_n}))_{n \geqslant 1}$ such that

$$\pi^1((z_n, \overrightarrow{u_n})) \in V_n, \quad \lim_{n \to +\infty} \pi^1((z_n, \overrightarrow{u_n})) = \pi^1((z, \overrightarrow{u})),$$

$$\lim_{n \to +\infty} (z_n, \overrightarrow{u_n}) = (z, \overrightarrow{u}) \quad \text{and}$$

$$\lim_{n \to +\infty} h_{t_n}(\pi^1((z_n, \overrightarrow{u_n}))) = \pi^1((z, \overrightarrow{u})).$$

The last of these limits when considered on $T^1\mathbb{H}$ implies the existence of a sequence $(\gamma_n)_{n \geqslant 1}$ in Γ such that $\lim_{n \to +\infty} \widetilde{h}_{t_n} \gamma_n((z_n, \overrightarrow{u_n})) = (z, \overrightarrow{u})$. Set $\widetilde{h}_{t_n}((z_n, \overrightarrow{u_n})) = (z'_n, \overrightarrow{u'_n})$. We have

$$\lim_{n \to +\infty} z'_n = u(+\infty) \quad \text{and} \quad \lim_{n \to +\infty} d(z'_n, \gamma_n^{-1}(z)) = 0.$$

This implies that $\lim_{n \to +\infty} \gamma_n^{-1}(z) = u(+\infty)$ and hence that $u(+\infty)$ is in $L(\Gamma)$.

Consider an element in $F(\Gamma)$ whose horocyclic trajectory is dense in $F(\Gamma)$. By Proposition 2.10, such a point exists. Moreover such point is non-wandering with respect to the horocycle flow. Since the non-wandering set is closed and invariant with respect to the horocycle flow, it contains $F(\Gamma)$. $\qquad\square$

3 Dense and periodic trajectories

Assume that Γ is a non-elementary Fuchsian group. Our motivation is to characterize the topology of a trajectory $h_\mathbb{R}(\pi^1((z, \overrightarrow{u})))$ included in $\Omega_h(T^1 S)$ in terms of the properties of the point $u(+\infty)$.

3.1 Dense horocyclic trajectories

If Γ is not elementary, Propositions 2.11 and 2.10 together imply that the set $\Omega_h(T^1 S)$ contains some dense horocyclic trajectories.

Theorem 3.1. *Let* $(z, \overrightarrow{u})$ *be in* $T^1 \mathbb{H}$. *The following are equivalent:*

(i) *the point* $u(+\infty)$ *is a horocyclic point in* $L(\Gamma)$;
(ii) *the orbit* $h_{\mathbb{R}}(\pi^1((z, \overrightarrow{u})))$ *is dense in* $\Omega_h(T^1 S)$.

To prove this theorem, we use the vectorial point of view and the following lemma, which follows from the definition of a horocyclic point and Property 1.9(i).

Lemma 3.2. *Let* $(z, \overrightarrow{u}) \in T^1 \mathbb{H}$. *The point* $u(+\infty)$ *is horocyclic if and only if there exists a sequence* $(M_n)_{n \geqslant 1}$ *in* M_Γ *such that* $\lim_{n \to +\infty} \| M_n(\mathrm{vect}((z, \overrightarrow{u}))) \| = 0$.

Proof (of Theorem 3.1). To prove Theorem 3.1, it is sufficient (by Proposition 2.1 and Lemma 3.2) to prove that the following statements are equivalent:

(i′) There exists a sequence $(M_n)_{n \geqslant 1}$ in M_Γ such that

$$\lim_{n \to +\infty} \| M_n(\mathrm{vect}((z, \overrightarrow{u}))) \| = 0.$$

(ii′) $\overline{M_\Gamma \, \mathrm{vect}((z, \overrightarrow{u}))} = E(\Gamma)$.

It is clear that (ii′) implies (i′).

Let us prove (i′) $\Rightarrow$ (ii′). We begin with the case in which $u(+\infty)$ is fixed by a hyperbolic isometry $\gamma \in \Gamma$. Let M be the element of M_Γ associated with γ. We can assume that $M(\mathrm{vect}((z, \overrightarrow{u}))) = \pm \lambda \, \mathrm{vect}((z, \overrightarrow{u}))$, with $0 < \lambda < 1$. By Proposition 2.6, there exists $(z', \overrightarrow{u'})$ in $T^1 \mathbb{H}$ such that $\overline{M_\Gamma \, \mathrm{vect}((z', \overrightarrow{u'}))} = E(\Gamma)$. If we prove that $\overline{M_\Gamma \, \mathrm{vect}((z, \overrightarrow{u}))}$ contains an element of $\pm \mathbb{R}^* \, \mathrm{vect}((z', \overrightarrow{u'}))$, then $\overline{M_\Gamma \, \mathrm{vect}((z, \overrightarrow{u}))} = E(\Gamma)$.

The point $u'(+\infty)$ belongs to $L(\Gamma)$, which is minimal, thus there exists $(\gamma_n)_{n \geqslant 1}$ in Γ such that $\lim_{n \to +\infty} \gamma_n(u(+\infty)) = u'(+\infty)$. Let M_n denote the element of M_Γ associated with γ_n and $(p_n)_{n \geqslant 1}$ be a sequence in $\mathbb{Z}$ such that $(\lambda^{p_n} \| M_n(\mathrm{vect}((z, \overrightarrow{u}))) \|)_{n \geqslant 1}$ converges to a real number $\alpha \neq 0$. Since $M_n M^{p_n}(\mathrm{vect}((z, \overrightarrow{u}))) = \pm \lambda^{p_n} M_n(\mathrm{vect}((z, \overrightarrow{u})))$ and $\gamma_n \gamma^{p_n} u(+\infty) = \gamma_n u(+\infty)$, one has

$$\lim_{n \to +\infty} M_n M^{p_n} \, \mathrm{vect}((z, \overrightarrow{u})) = \pm \alpha \frac{\mathrm{vect}((z', \overrightarrow{u'}))}{\| \mathrm{vect}((z', \overrightarrow{u'})) \|}.$$

Assume now that $u(+\infty)$ is horocyclic and is not fixed by any hyperbolic isometry in Γ. Let us show that $\overline{M_\Gamma \, \mathrm{vect}((z, \overrightarrow{u}))}$ contains an eigenvector, modulo ± 1, of an element of M_Γ associated to a hyperbolic isometry in Γ.

This will prove that M_Γ vect$((z, \vec{u}))$ is dense in $E(\Gamma)$. Let γ be a hyperbolic isometry in Γ. Denote M the element of M_Γ associated with γ. The sequence $(\gamma^{-n}u(+\infty))_{n\geqslant 1}$ converges to γ^-. Consider a sequence $(M_n)_{n\geqslant 1}$ in M_Γ such that $\lim_{n\to+\infty} \|M_n(\text{vect}((z, \vec{u})))\| = 0$. Take a vector w in $(\mathbb{R}^*)^2$ which projects to vect$((z, \vec{u}))$ and $\widehat{M}, \widehat{M}_n$ in SL$(2, \mathbb{R})$ which project to M and M_n respectively. Let w^+, w^- be eigenvectors of $\widehat{M}$ such that

$$\widehat{M}w^+ = \lambda w^+ \quad \text{and} \quad \widehat{M}w^- = \frac{1}{\lambda}w^-,$$

with $|\lambda| > 1$. Set $\widehat{M}_n(w) = a_n w^+ + b_n w^-$. Then

$$\lim_{n\to+\infty} (a_n^2 + b_n^2) = 0.$$

Furthermore,

$$\widehat{M}^{-n}\widehat{M}_n(w) = (a_n/\lambda^n)w^+ + b_n\lambda^n w^-,$$

hence, after passing to a subsequence, one can assume that $(\widehat{M}^{-n}\widehat{M}_n(w))_{n\geqslant 1}$ converges to βw^- with $\beta \neq 0$. It follows that $(M^{-n}M_n(\text{vect}((z, \vec{u}))))_{n\geqslant 1}$ converges to an eigenvector, modulo ± 1, of a matrix associated to a hyperbolic isometry of Γ. $\square$

As an application of this theorem, we prove Theorem III.4.3 about the topological mixing property of the geodesic flow on $\Omega_g(T^1S)$.

3.2 Relationship between $h_\mathbb{R}$ and $g_\mathbb{R}$, and application

The horocycle flow is closely related to geodesic flow. Namely we have

Property 3.3. *Let $(z, \vec{u})$ be in $T^1\mathbb{H}$ and let s, t be in $\mathbb{R}$. One has*

$$\widetilde{g}_t \circ \widetilde{h}_s \circ \widetilde{g}_{-t}((z, \vec{u})) = \widetilde{h}_{se^{-t}}((z, \vec{u})).$$

Proof. Because the action of the group G of positive isometries of $\mathbb{H}$ commutes with those of $\widetilde{g}_\mathbb{R}$ and $\widetilde{h}_\mathbb{R}$, and because this group acts transitively on $T^1\mathbb{H}$, it is sufficient to prove this relationship when $z = i$ and $u(+\infty) = \infty$.

In this case, the image by $\widetilde{g}_{-t}$ of $(i, \vec{u})$ is $(e^{-t}i, \vec{v})$, with $v(+\infty) = \infty$. It follows that $\widetilde{h}_s\widetilde{g}_{-t}(i, \vec{u}) = (e^{-t}i + se^{-t}, \vec{u}')$, with $u'(+\infty) = \infty$ (Fig. V.6). Hence

$$\widetilde{g}_t(\widetilde{h}_s(\widetilde{g}_{-t}((i, \vec{u})))) = (i + e^{-t}s, \vec{u''}),$$

with $u''(+\infty) = \infty$.

We conclude using the fact that $h_{se^{-t}}((i, \vec{u})) = (i + e^{-t}s, \vec{u''})$. $\square$

As application of this relationship, we use properties of the horocyclic flow on T^1S to obtain Theorem III.4.3 relative to the behavior of the geodesic flow.

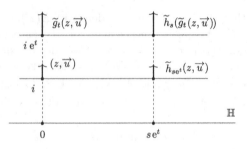

Fig. V.6.

Proof of Theorem III.4.3

Recall the statement of this theorem:

Let O and V be open and non-empty subsets of $\Omega_g(T^1S)$. There exists $T > 0$ such that for all $t \geqslant T$, $g_t(O) \cap V \neq \varnothing$.

Proof. We prove this statement by contradiction. Assume that there are two non-empty open subsets O and V of $\Omega_g(T^1S)$ and an unbounded sequence $(t_n)_{n \geqslant 1}$ such that $O \cap g_{t_n}(V) = \varnothing$. We can suppose that $\lim_{n \to +\infty} t_n = -\infty$. By Theorem III.3.3, there exists $\pi^1((z, \overrightarrow{u}))$ in V which is periodic with respect to $g_{\mathbb{R}}$. Let T denote its period and define $t_n = r_n T + s_n$ where $-r_n \in \mathbb{N}$ and $-T < s_n \leqslant 0$. After passing to a subsequence, one can assume that $(s_n)_{n \geqslant 1}$ converges to a real number s. The point $u(+\infty)$ is horocyclic. Thus by Theorem 3.1, one has $\overline{h_{\mathbb{R}}(\pi^1((z, \overrightarrow{u})))} = \Omega_h(T^1S)$. By Proposition 2.11 and Theorem III.2.1, we have $\Omega_g(T^1S) \subset \Omega_h(T^1S)$. Hence there exists t in $\mathbb{R}$ such that $h_t(\pi^1((z, \overrightarrow{u})))$ belongs to $g_{-s}(O)$. Consider the hyperbolic isometry γ in Γ such that $u(+\infty) = \gamma^+$ and $\ell(\gamma) = T$. One has $\gamma^n((z, \overrightarrow{u})) = \widetilde{g}_{nT}((z, \overrightarrow{u}))$. Using this relationship and Property 3.3, one obtains the equality $\gamma^{-r_n} \widetilde{g}_{-r_n T}(\widetilde{h}_t((z, \overrightarrow{u}))) = \widetilde{h}_{t e^{r_n T}} \circ \widetilde{g}_{-2r_n T}((z, \overrightarrow{u}))$. This equality implies that $(g_{-r_n T}(h_t(\pi^1((z, \overrightarrow{u})))))_{n \geqslant 1}$ converges to $\pi^1((z, \overrightarrow{u}))$ and thus that for large enough n, $g_{s+r_n T}V \cap O \neq \varnothing$. Since $\lim_{n \to +\infty}(s_n - s) = 0$, it follows that for large enough n, we have $g_{t_n}V \cap O \neq \varnothing$. This contradicts our initial hypothesis. $\qquad \square$

3.3 Periodic horocyclic trajectories and their periods

We now focus on the existence of periodic elements of the flow $h_{\mathbb{R}}$ on T^1S. By definition, if $\pi^1((z, \overrightarrow{u}))$ is periodic, there exist $T > 0$ and $\gamma \in \Gamma$ such that

$$\widetilde{h}_T((z, \overrightarrow{u})) = \gamma((z, \overrightarrow{u})).$$

Proposition 3.4. *Let $(z, \overrightarrow{u})$ be in $T^1\mathbb{H}$ and let γ be a positive isometry of $\mathbb{H}$. The following are equivalent:*

(i) *there exists $T > 0$ such that $\widetilde{h}_T((z, \overrightarrow{u})) = \gamma((z, \overrightarrow{u}))$;*
(ii) *the isometry γ is parabolic and fixes $u(+\infty)$.*

Proof.

(i) $\Rightarrow$ (ii). Suppose that there exist $T > 0$ and γ such that $\widetilde{h}_T((z, \overrightarrow{u})) = \gamma((z, \overrightarrow{u}))$. For all n in $\mathbb{Z}^*$, one has $\widetilde{h}_{nT}((z, \overrightarrow{u})) = \gamma^n((z, \overrightarrow{u}))$. Hence γ fixes $u(+\infty)$. Furthermore $\gamma \neq \mathrm{Id}$ and γ is not hyperbolic since $\lim_{n \to +\infty} \gamma^{-n}(z) = \lim_{n \to +\infty} \gamma^n(z)$. Therefore γ is parabolic.

(ii) $\Rightarrow$ (i). Let γ be a parabolic isometry and let $(z, \overrightarrow{u})$ be an element of $T^1 \mathbb{H}$ satisfying $\gamma(u(+\infty)) = u(+\infty)$. This isometry preserves the oriented horocycle associated with $(z, \overrightarrow{u})$ (see Sect. I.2.2). Hence there exists $T > 0$ such that $\gamma((z, \overrightarrow{u})) = \widetilde{h}_T((z, \overrightarrow{u}))$ (Fig. V.7). $\square$

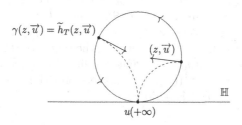

$$\gamma(z, \overrightarrow{u}) = \widetilde{h}_T(z, \overrightarrow{u})$$

$(z, \overrightarrow{u})$

$\mathbb{H}$

$u(+\infty)$

Fig. V.7.

Clearly, one obtains:

Corollary 3.5.

(i) *The element $\pi^1((z, \overrightarrow{u}))$ is periodic with respect to the horocycle flow on $\Omega_h(T^1 S)$ if and only if $u(+\infty)$ is fixed by a parabolic isometry of Γ.*

(ii) *The set $\Omega_h(T^1 S)$ contains periodic trajectories if and only if the group Γ contains parabolic isometries.*

For example, if Γ is a Schottky group generated by two hyperbolic isometries, then $\Omega_h(T^1 S)$ does not contain any periodic trajectory. On the other hand, such trajectories do exist if Γ is the modular group.

Note that, if $\pi^1((z, \overrightarrow{u}))$ is periodic with respect to $h_{\mathbb{R}}$, then $\pi^1(\widetilde{g}_t((z, \overrightarrow{u})))$ is likewise periodic for all $t \in \mathbb{R}$. It follows that, if it is not empty, the set of periodic trajectories of $h_{\mathbb{R}}$ contains a subset which is in one-to-one correspondence with $\mathbb{R}$.

As in the case of the geodesic flow, if $\pi^1((z, \overrightarrow{u}))$ is a periodic element with respect to the flow $h_{\mathbb{R}}$, we write T_u to denote its period. Let γ be the parabolic isometry of Γ satisfying

$$\widetilde{h}_{T_u}((z, \overrightarrow{u})) = \gamma((z, \overrightarrow{u})).$$

Does γ determine the period T_u? Unlike the case of the geodesic flow (see Sect. III.2), the answer is "No." To see this, suppose that $u(+\infty) = \infty$. Under this hypothesis, γ is necessarily a translation $t(z) = z + a$ and therefore

$$T_u = \frac{|a|}{\operatorname{Im} z}.$$

This equality shows that T_u does not depend only on the translation t, and that the set of all periods of periodic elements of $\Omega_h(T^1 S)$ is $\mathbb{R}_+^*$.

Proposition 3.6. *If the set of periodic elements of the horocycle flow is non-empty, then it is dense in $\Omega_h(T^1 S)$.*

Proof. Suppose that there exists a periodic element $\pi^1((z_0, \overrightarrow{u_0}))$ in $T^1 S$. Let $\pi^1((z, \overrightarrow{u}))$ be in $\Omega_h(T^1 S)$. Since $L(\Gamma)$ is a minimal set with respect to the action of Γ, there exists $(\gamma_n)_{n \geqslant 1}$ in Γ such that $\lim_{n \to +\infty} \gamma_n(u_0(+\infty)) = u(+\infty)$. Consider the sequence $((z, \overrightarrow{u_n}))_{n \geqslant 1}$ in $T^1 \mathbb{H}$ defined by $u_n(+\infty) = \gamma_n(u_0(+\infty))$. One has $\lim_{n \to +\infty}(z, \overrightarrow{u_n}) = (z, \overrightarrow{u})$, hence $\lim_{n \to +\infty} \pi^1((z, \overrightarrow{u_n})) = \pi^1((z, \overrightarrow{u}))$. Furthermore, the point $\gamma_n(u(+\infty))$ is parabolic, thus $\pi^1((z, \overrightarrow{u_n}))$ is periodic. $\qquad\square$

4 Characterization of geometrically finite Fuchsian groups

Recall that a Fuchsian group is geometrically finite if and only if every point in its limit set is either horocyclic or parabolic (see Theorem I.4.13).

The geometric finiteness of a group can be characterized using the topological dynamics of $h_{\mathbb{R}}$. Actually, the following theorem stems directly from Theorem 3.1 and from Proposition 3.5.

Theorem 4.1. *A non-elementary Fuchsian group is geometrically finite if and only if the trajectories of $h_{\mathbb{R}}$ restricted to $\Omega_h(T^1 S)$ are dense in $\Omega_h(T^1 S)$ or periodic.*

For example, if Γ is a Schottky group generated by two hyperbolic isometries, all trajectories of $h_{\mathbb{R}}$ are dense in $\Omega_h(T^1 S)$. If Γ is the modular group, the horocyclic trajectories are dense in $T^1 S$ or they are periodic.

Among the geometrically finite Fuchsian groups, recall that convex-cocompact groups are those whose limit sets are entirely composed of horocyclic points (Corollary I.4.17). If Γ is such a group, all trajectories of $h_{\mathbb{R}}$ are dense in $\Omega_h(T^1 S)$; in other words, $\Omega_h(T^1 S)$ is a minimal set with respect to the flow $h_{\mathbb{R}}$.

Conversely, if all trajectories of $h_{\mathbb{R}}$ are dense in $\Omega_h(T^1 S)$, then all points of $L(\Gamma)$ are horocyclic and therefore Γ is convex-cocompact.

One can state the following properties:

Proposition 4.2. *Let Γ be a non-elementary Fuchsian group.*

(i) *The group Γ is convex-cocompact if and only if $\Omega_h(T^1S)$ is a minimal set with respect to the flow $h_{\mathbb{R}}$.*

(ii) *The group Γ is a uniform lattice if and only if T^1S is a minimal set with respect to the horocycle flow on T^1S.*

Using Proposition 2.1, one can translate Propositions 4.1 and 4.2 into vectorial terms. We thus obtain the following characterization of geometric finiteness and of convex-cocompactness in terms of the dynamics of the group M_Γ on $E(\Gamma)$.

Proposition 4.3. *Let Γ be a non-elementary Fuchsian group.*

(i) *The group Γ is geometrically finite if and only if for all vector $v \in E(\Gamma)$, either $\overline{M_\Gamma(v)} = E(\Gamma)$ or there exists $M \in M_\Gamma - \{\mathrm{Id}\}$ such that $Mv = v$.*

(ii) *The group Γ is convex-cocompact if and only if $\overline{M_\Gamma(v)} = E(\Gamma)$ for all $v \in E(\Gamma)$.*

(iii) *The group Γ is a uniform lattice if and only if $\overline{M_\Gamma(v)} = E$ for all $v \in E$.*

In the case of a Schottky group generated by two hyperbolic isometries, every orbit of M_Γ has dense restriction to $E(\Gamma)$. If Γ is the modular group, one deduces from Proposition 4.3 the following result:

Property 4.4. *Let $\left(\begin{smallmatrix} x \\ y \end{smallmatrix}\right) \in \mathbb{R}^2$. If $y \neq 0$ and if $x/y \notin \mathbb{Q}$, then $\overline{\mathrm{SL}(2,\mathbb{Z})\left(\begin{smallmatrix} x \\ y \end{smallmatrix}\right)} = \mathbb{R}^2$.*

Exercise 4.5. Prove Property 4.4.

In the case of the modular group, if a vector w in E is fixed by a non-trivial element of M_Γ then $w = \lambda\left(\begin{smallmatrix} p \\ q \end{smallmatrix}\right)$ for some $(p,q) \in \mathbb{Z} \times \mathbb{Z}$. Thus $M_\Gamma(v)$ is a discrete set in E. This is a general phenomenon as described in the following exercise.

Exercise 4.6. Let Γ be a non-elementary Fuchsian group and let w in $E(\Gamma)$ be fixed by a non-trivial element of M_Γ. Prove that the orbit $M_\Gamma(v)$ is a discrete subset of $E(\Gamma)$.
(Hint: use Corollary 2.2.)

5 Comments

The horocycle flow is closely related to the geodesic flow. The collective behavior of the geodesic trajectories is reflected by the horocycle flow in the sense of Property 3.3: two elements $\pi^1((z, \overrightarrow{u}))$ and $\pi^1((z', \overrightarrow{u'}))$ in the quotient $\Gamma\backslash T^1\mathbb{H}$ belong to the same horocyclic trajectory if and only if the distance between $g_t(\pi^1((z, \overrightarrow{u})))$ and $g_t(\pi^1((z', \overrightarrow{u'})))$ converges to 0 when t tends to $+\infty$.

Although the behavior of individual geodesics is very unpredictable as shown in Chaps. III and IV, when Γ is geometrically finite, the collective behavior of the geodesic flow is regular (Proposition 4.2). If the hypothesis of geometric finiteness is not satisfied, the horocyclic flow can be very irregular, as illustrated in examples constructed by M. Kulikov [43]. In these examples, a group Γ is devised for which the horocycle flow on $\Gamma \backslash T^1 \mathbb{H}$ does not admit any minimal set.

The relationship between these two flows allows one, for example, to prove the topological mixing of $g_{\mathbb{R}}$ (Theorem III.4.3). This relationship is especially useful when tackling metric questions, the historical example being the proof due to G. Hedlund [32] in the setting of lattices, of the ergodicity of $h_{\mathbb{R}}$ with respect to the Liouville measure.

In the general case where X is a pinched Hadamard manifold and Γ is a non-elementary Kleinian group acting on X, the notion of horocycle flow on $T^1 X$ only makes sense if the dimension of X is equal to 2. If X is not a surface, this notion is replaced by that of the *strong stable foliation* on $T^1 X$ which generalizes the foliation of $T^1 \mathbb{H}$ by the trajectories of $\tilde{h}_{\mathbb{R}}$ [5, 27, 54]. The projection of the leaves of this foliation to X are the horospheres of X. The part of the non-wandering set of the horocycle flow on $\Gamma \backslash T^1 \mathbb{H}$ is then played by $\Omega_h(\Gamma \backslash T^1 X)$, the set obtained by projecting to $\Gamma \backslash T^1 X$ the leaves of $T^1 X$ corresponding to horospheres centered at points of $L(\Gamma)$. In this general setting, the existence of a leaf which is dense in $\Omega_h(\Gamma \backslash T^1 X)$ is an open question. This question is equivalent to that of the topological mixing of the geodesic flow on $\Omega_g(\Gamma \backslash T^1 X)$ as well as that of the density of the length spectrum [19]. In the context of the Poincaré half plane, we have proved this existence (Corollary 2.10) using a vectorial point of view. This approach is found in [16]. Adding the assumption that there exists a dense leaf in $\Omega_h(\Gamma \backslash T^1 X)$, most of the results of this chapter can be generalized [5, 19, 27, 54].

As with the geodesic flow, we have not addressed the metric aspect of the horocycle flow. The texts of É. Ghys [32] and S. Starkov [59] provide a good introduction. Let us outline the basics.

Consider the horocycle flow $\tilde{h}_{\mathbb{R}}$ on $T^1 \mathbb{D}$ where $\mathbb{D}$ is the Poincaré disk. Let 0 be the center of this disk. Each trajectory is identified with a pair (x, s) where $x \in \mathbb{D}(\infty)$, $s \in \mathbb{R}$ and $s = B_x(0, z)$, for points z on the horocycle associated with this trajectory.

Let Γ be a non-elementary Fuchsian group. This identification allows one to construct a measure N on $T^1 \mathbb{H}$ which is invariant with respect to this flow and with respect to Γ, defined by

$$N(dx\, ds) = e^{s\delta(\Gamma)} m(dx)\, ds,$$

where m is a Patterson measure on $L(\Gamma)$ and $\delta(\Gamma)$ is the critical exponent of the Poincaré series associated with Γ (see the Comments of Chap. I).

The measure N induces another measure $\overline{N}$ on $\Gamma \backslash T^1 \mathbb{H}$ which is invariant with respect to the horocycle flow and whose support is $\Omega_h(\Gamma \backslash T^1 \mathbb{H})$ [5, 54]. If Γ is a lattice, this measure is finite. Otherwise it is infinite. Under the assumption that Γ is geometrically finite, the horocycle flow is ergodic with respect to this measure [32, 53, 54].

In the general case of a pinched Hadamard manifold of arbitrary dimension, the construction of the measure $\overline{N}$ and its resulting ergodicity can be generalized, provided that one again assumes the existence of a leaf which is dense in $\Omega_h(\Gamma\backslash T^1X)$ [54]. Adding a further assumption of finiteness on the Patterson-Sullivan measures on $\Gamma\backslash T^1X$ (see the Comments from Chap. III), one obtains a classification of ergodic measures which are invariant with respect to the strong stable foliation ([47, 53] and [54, Theorem 6.5]).

VI

The Lorentzian point of view

In the previous chapter (Sect. V.2), we established a correspondence between the dynamics of the horocycle flow on $\Gamma \backslash T^1\mathbb{H}$ and the dynamics of the linear group associated with Γ on $\{\pm\,\mathrm{Id}\}\backslash\mathbb{R}^2 - \{0\}$.

Our motivation in this chapter, is to construct a linear representation of Γ taking into account simultaneously the dynamics of the horocycle and of the geodesic flows. Many proofs are reformulations of proofs given in the previous chapters. In this case, they are left to the reader. Appendix B can be useful in this chapter.

To this end, we work in the space $\mathbb{R}^3$ equipped with the *Lorentz* bilinear form

$$b(x, x') = x_1 x_1' + x_2 x_2' - x_3 x_3'.$$

Each real number t is associated with a surface

$$\mathcal{H}_t = \{x \in \mathbb{R}^3 \mid b(x, x) = t\}.$$

If t is strictly negative, $\mathcal{H}_t$ is a *hyperboloid* of two sheets (Fig. VI.1). In this

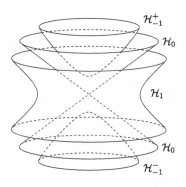

Fig. VI.1.

F. Dal'Bo, *Geodesic and Horocyclic Trajectories*, Universitext,
DOI 10.1007/978-0-85729-073-1_6, © Springer-Verlag London Limited 2011

case, one sets

$$\mathcal{H}_t^+ = \mathcal{H}_t \cap (\mathbb{R}^2 \times \mathbb{R}^+) \quad \text{and} \quad \mathcal{H}_t^- = \mathcal{H}_t \cap (\mathbb{R}^2 \times \mathbb{R}^-).$$

If t is strictly positive, $\mathcal{H}_t$ is connected and is a *hyperboloid* of one sheet (Fig. VI.1).

Finally, $\mathcal{H}_0$ is the *light cone*, an object associated with special relativity (Fig. VI.1). This cone minus the point $(0,0,0)$ has two connected components

$$\mathcal{H}_0^{+*} = \mathcal{H}_0 \cap \mathbb{R}^2 \times \mathbb{R}_+^* \quad \text{and} \quad \mathcal{H}_0^{-*} = \mathcal{H}_0 \cap \mathbb{R}^2 \times \mathbb{R}_-^*.$$

The group $O(2,1)$ of orthogonal transformations of b acts on each surface $\mathcal{H}_t$. This group is not connected. Let $O^0(2,1)$ denote the connected component of $O^0(2,1)$ containing the identity.

Throughout this chapter, $\|x\|$ will indicate the Euclidean norm of x in $\mathbb{R}^3$.

1 The hyperboloid model

In our present context, the Poincaré disk is considered to be the subset of $\mathbb{R}^3$ defined by

$$\mathbb{D} = \{x \in \mathbb{R}^2 \times \{0\} \mid \|x\| < 1\},$$

equipped with the metric g as defined in Sect. I.1.5.

The purpose of this section is to construct a Riemannian structure on the sheet $\mathcal{H}_{-1}^+$, for which $\mathcal{H}_{-1}^+$ is isometric to the Poincaré disk, and to understand its geometry from a Lorentzian point of view.

1.1 Construction of the metric and compactification

Consider the stereographic projection $P : \mathcal{H}_{-1}^+ \to \mathbb{D}$, which sends x to the point $P(x)$ which is the intersection of $\mathbb{D}$ with the line passing through x and through the point $(0,0,-1)$ (Fig. VI.2). This map is a diffeomorphism whose analytic expression is

$$P(x_1, x_2, t) = \left(\frac{x_1}{1+t}, \frac{x_2}{1+t}, 0 \right).$$

Let g^{L} denote the metric on $\mathcal{H}_{-1}^+$ obtained by pulling back the hyperbolic metric g on $\mathbb{D}$ by P^{-1}. By definition, for $x \in \mathcal{H}_{-1}^+$ and $\overrightarrow{v}, \overrightarrow{v}' \in T_x^1 \mathcal{H}_{-1}^+$, we have

$$g_x^{\mathrm{L}}(\overrightarrow{v}, \overrightarrow{v}') = \frac{4}{(1 - \|P(x)\|^2)^2} \langle T_x P(\overrightarrow{v}), T_x P(\overrightarrow{v}') \rangle.$$

Exercise 1.1. Prove the equality

$$g_x^{\mathrm{L}}(\overrightarrow{v}, \overrightarrow{v}') = b(\overrightarrow{v}, \overrightarrow{v}').$$

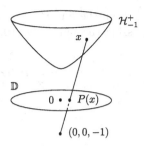

Fig. VI.2.

We deduce from Exercise 1.1, that on each tangent plane of $\mathcal{H}^+_{-1}$, the metric g^L corresponds to the restriction of b to that tangent plane. Equipped with this metric, $\mathcal{H}^+_{-1}$ is isometric to the Poincaré disk and therefore to the Poincaré half-plane.

Exercise 1.2. Recall that G is the group of positive isometries of $(\mathbb{D}, g)$.

(i) Prove that $\mathcal{H}^+_{-1}$ is invariant with respect to the group $O^0(2,1)$.
(ii) Prove the equality
$$P O^0(2,1)P^{-1} = G.$$

It follows from Exercise 1.2 that the group $O^0(2,1)$ is the group of orientation-preserving isometries of $(\mathcal{H}^+_{-1}, g^L)$. The action of this group is transitive on $\mathcal{H}^+_{-1}$ and simply transitive on the unitary tangent bundle $T^1\mathcal{H}^+_{-1}$ of $\mathcal{H}^+_{-1}$ (Property I.2.3).

Exercise 1.3. Prove that the geodesics of $(\mathcal{H}^+_{-1}, g^L)$ correspond to the intersections of $\mathcal{H}^+_{-1}$ with planes passing through the point $(0,0,0)$ and through a point of $\mathcal{H}^+_{-1}$ (Fig. VI.3).

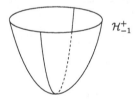

Fig. VI.3.

We compactify $(\mathcal{H}^+_{-1}, g^L)$ by applying P^{-1} to the compactification of $(\mathbb{D}, g)$. To accomplish this, consider the space $\mathcal{H}^+_{-1}(\infty)$ of lines in the cone $\mathcal{H}_0$. The bijection P^{-1} extends to a bijection, again denoted P^{-1}, of $\mathbb{D} \cup \mathbb{D}(\infty)$ on $\mathcal{H}^+_{-1} \cup \mathcal{H}^+_{-1}(\infty)$ defined on $\mathbb{D}(\infty)$ by

$$P^{-1}((\cos\sigma, \sin\sigma, 0)) = \{(t\cos\sigma, t\sin\sigma, t) \mid t \in \mathbb{R}\}.$$

In other words, if $x \in \mathbb{D}(\infty)$, then the line $P^{-1}(x)$ contains the point $(0,0,0)$ and is parallel to the line passing through x and $(0,0,-1)$ (Fig. VI.4).

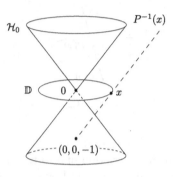

Fig. VI.4.

The set $\mathcal{H}_{-1}^+(\infty)$ is called the boundary at infinity of $\mathcal{H}_{-1}^+$. Let us equip $\mathcal{H}_{-1}^+ \cup \mathcal{H}_{-1}^+(\infty)$ with the topology induced by P in which a neighborhood of a point y is the pre-image of a neighborhood of $P(y)$. In this topology, $\mathcal{H}_{-1}^+ \cup \mathcal{H}_{-1}^+(\infty)$ is compact and P is a homeomorphism. More explicitly, the convergence of an unbounded sequence $(y_n)_{n\geqslant 1}$ in $\mathcal{H}_{-1}^+$ to a line D in $\mathcal{H}_{-1}^+(\infty)$ corresponds to the Euclidean convergence to D of the sequence of lines passing through the origin and y_n.

The endpoints of a geodesic in $\mathcal{H}_{-1}^+$ correspond to two lines in the cone $\mathcal{H}_0$ contained in the plane passing through this geodesic and the point $(0,0,0)$ (Fig. VI.5).

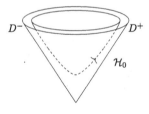

Fig. VI.5.

Since the action of G on $\mathbb{D}$ extends to $\mathbb{D}(\infty)$ via an action of homeomorphisms, the action of $O^0(2,1)$ extended similarly to $\mathcal{H}_{-1}^+ \cup \mathcal{H}_{-1}^+(\infty)$.

Exercise 1.4. Prove that the action of $O^0(2,1)$ on $\mathcal{H}_{-1}^+(\infty)$, which is P-conjugate to the action of G on $\mathbb{D}(\infty)$, corresponds to the projective action of $O^0(2,1)$ on the space of lines in $\mathcal{H}_0$.

1.2 Classification of positive isometries and Busemann cocycles

Let f be a non-trivial transformation of $O^0(2,1)$. One says that f is respectively *elliptic*, *parabolic* or *hyperbolic* if PfP^{-1} is (see Sect. I.2). Translated in terms of isotropic eigenvectors (i.e., $b(\vec{v}, \vec{v}) = 0$ and $\vec{v} \neq \vec{0}$), this classification boils down to the following property:

- either f does not admit any isotropic eigenvector (f is elliptic);
- or f admits exactly one isotropic eigendirection (f is parabolic);
- or f admits two distinct isotropic eigendirections (f is hyperbolic).

Let A denote the subgroup of elements of $O^0(2,1)$ globally fixing the lines $D_0 = \mathbb{R}(1,0,1)$ and $D_1 = \mathbb{R}(-1,0,1)$. Let N denote the subgroup of elements of $O^0(2,1)$ globally fixing D_0. Let K denote the subgroup of elements fixing the point $x_0 = (0,0,1)$.

For all $t \in \mathbb{R}$, set

$$a_t = \begin{pmatrix} \cosh t & 0 & \sinh t \\ 0 & 1 & 0 \\ \sinh t & 0 & \cosh t \end{pmatrix}, \quad n_t = \begin{pmatrix} 1 - t^2/2 & t & t^2/2 \\ -t & 1 & t \\ -t^2/2 & t & t^2/2 + 1 \end{pmatrix},$$

$$k_t = \begin{pmatrix} \cos t & -\sin t & 0 \\ \sin t & \cos t & 0 \\ 0 & 0 & 1 \end{pmatrix}.$$

Exercise 1.5. Prove the following properties

(i) $A = \{a_t/t \in \mathbb{R}\}, \quad N = \{n_t \mid t \in \mathbb{R}\}, \quad K = \{k_t \mid t \in [0, 2\pi)\}$.
(ii) For all f in $O^0(2,1)$, there exist t, t', t'' and s, s', s'' such that

$$f = k_t a_{t'} n_{t''} \quad \text{(see Proposition I.2.4)},$$
$$f = k_s a_{s'} k_{s''} \quad \text{(see Proposition I.2.4)}.$$

The notion of the Busemann cocycle makes sense on $\mathcal{H}^+_{-1}$. To see this, let D be an element of $\mathcal{H}^+_{-1}(\infty)$, and let x, y be two points in $\mathcal{H}^+_{-1}$. Consider the arclength parametrization $(R(t))_{t \geqslant 0}$ of the geodesic ray beginning at x and ending at D and set

$$F(t) = d^L(x, R(t)) - d^L(y, R(t)),$$

where d^L is the distance function induced on $\mathcal{H}^+_{-1}$ by g^L (see Sect. I.1).
By construction of the metric g^L, one has

$$F(t) = d(P(x), r(t)) - d(P(y), r(t)),$$

where $(r(t))_{t \geqslant 0}$ is the arclength parametrization of the geodesic ray $[P(x), P(D))$. It follows that the limit of F as t goes to $+\infty$ exists and one has (see Sect. I.1)

$$B_{P(D)}(P(x), P(y)) = \lim_{t \to +\infty} F(t).$$

Thus the *Busemann cocycle* centered at D, calculated at x and y, is defined as follows:
$$B_D(x,y) = B_{P(D)}(P(x), P(y)).$$

A horocycle of $\mathcal{H}^+_{-1}$ centered at $D \in \mathcal{H}^+_{-1}(\infty)$ is by definition a level set of the function

$$\mathcal{H}^+_{-1} \longrightarrow \mathbb{R}$$
$$x \longmapsto B_D((0,0,1), x).$$

Clearly, the image by P of such a horocycle is a horocycle of $\mathbb{D}$. Moreover a horocycle of $\mathcal{H}^+_{-1}$ centered at D is invariant with respect to the group of parabolic isometries of $O^0(2,1)$ fixing D.

Exercise 1.6. Prove that a horocycle of $\mathcal{H}^+_{-1}$ centered at D passing through $x \in \mathcal{H}^+_{-1}$ is the intersection of $\mathcal{H}^+_{-1}$ with the plane passing through x which is parallel to the tangent plane of the cone $\mathcal{H}_0$ containing the line D (Fig. VI.6).

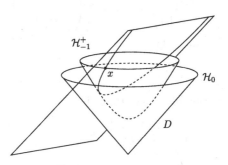

Fig. VI.6.

1.3 Lorentz groups and limit sets

By definition, a *Lorentz group* is a subgroup of $O^0(2,1)$ which is P-conjugate to a Fuchsian group; equivalently, such a group is a discrete subgroup of $O^0(2,1)$.

Let Γ_L be such a group and Γ_F its associated Fuchsian group. One has

$$\Gamma_F = P\Gamma_L P^{-1}.$$

The action of a Lorentz group on $\mathcal{H}^+_{-1}$ is properly discontinuous (Property I.2.9).

When Γ_F is infinite, we define its *limit set*, $L(\Gamma_F)$, as the intersection of $\mathcal{H}^+_{-1}(\infty)$ with the closure of any orbit $\Gamma_F(x)$, with $x \in \mathcal{H}^+_{-1}$. We have

$$L(\Gamma_F) = P(L(\Gamma_L)).$$

One says that Γ_L is *elementary* if Γ_F is.

As with $L(\Gamma_{\mathrm{F}})$, the notions of *horocyclic*, *conical* and *parabolic* points can be defined (see Sect. I.3). The following proposition gives a characterization of these points in terms of the linear action of Γ_{L}. For $f \in \mathrm{O}^0(2,1)$, we define the norm $\|f\| = \sup_{x \neq 0} \|f(x)\|/\|x\|$.

Proposition 1.7. *Let Γ_{L} be a Lorentz group. Take $D \in \mathcal{H}^+_{-1}(\infty)$ and let y be a direction vector for D.*

(i) *The line D is horocyclic with respect to Γ_{L} if and only if there exists a sequence $(\gamma_n)_{n \geqslant 1}$ in Γ_{L} such that $\lim_{n \to +\infty} \|\gamma_n y\| = 0$.*

(ii) *The line D is conical with respect to Γ_{L} if and only if there exists a sequence $(\gamma_n)_{n \geqslant 1}$ in Γ_{L} such that $\lim_{n \to +\infty} \|\gamma_n^{-1}\| = +\infty$ and such that the sequence $(\|\gamma_n^{-1}\|\|\gamma_n y\|)_{n \geqslant 1}$ is bounded.*

Proof. Set $D_0 = \mathbb{R} \left(\begin{smallmatrix} 1 \\ 0 \\ 1 \end{smallmatrix} \right)$, $y_0 = \left(\begin{smallmatrix} 1 \\ 0 \\ 1 \end{smallmatrix} \right)$ and $x_0 = \left(\begin{smallmatrix} 0 \\ 0 \\ 1 \end{smallmatrix} \right)$.

(i) Let us begin with the case in which $D = D_0$. By Exercise 1.5(ii), a transformation f in $\mathrm{O}^0(2,1)$ can be decomposed into $k_t a_{t'} n_{t''}$ with $k_t \in K$, $a_{t'} \in A$ and $n_{t''} \in N$. One has $B_{D_0}(x_0, f^{-1}(x_0)) = B_{D_0}(x_0, a_{t'}^{-1}(x_0))$, hence $B_{D_0}(x_0, f^{-1}(x_0)) = -t'$. Furthermore, $\|f(y_0)\| = \sqrt{2}e^{t'}$, hence $\|f(y_0)\| = \|y_0\|e^{-B_{D_0}(x_0, f^{-1}(x_0))}$.

Therefore, $\lim_{n \to +\infty} \|\gamma_n(y_0)\| = 0$ if and only if

$$\lim_{n \to +\infty} B_{D_0}(x_0, \gamma_n^{-1}(x_0)) = +\infty.$$

This proves the equivalence in (i) when $y = y_0$.

Suppose now that D is arbitrary in $\mathcal{H}^+_{-1}(\infty)$. Notice that the group $\mathrm{O}^0(2,1)$ acts transitively on $\mathcal{H}^{+*}_0$. Let f be in $\mathrm{O}^0(2,1)$ such that $y = f(y_0)$ is a direction of D. One has $\|\gamma_n(y)\| = \|\gamma_n f(y_0)\|$, hence $\lim_{n \to +\infty} \|\gamma_n(y)\| = 0$ if and only if $\lim_{n \to +\infty} B_{D_0}(x_0, f^{-1}\gamma_n^{-1}(x_0)) = +\infty$. The equivalence in part (i) can then be deduced from the relation

$$(*) \qquad B_{D_0}(x_0, f^{-1}\gamma_n^{-1}(x_0)) = B_D(f(x_0), x_0) + B_D(x_0, \gamma_n^{-1}(x_0)).$$

(ii) Again let us begin with the case where $D = D_0$. By Exercise 1.5(iii), a transformation f in $\mathrm{O}^0(2,1)$ can be decomposed into $k_t a_{t'} k_{t''}$ with $k_t, k_{t''} \in K$ and $a_{t'} \in A$. One has $d^{\mathrm{L}}(x_0, f(x_0)) = d^{\mathrm{L}}(x_0, a_{t'}(x_0))$ and $d^{\mathrm{L}}(x_0, a_{t'}(x_0)) = |t'|$. Furthermore, $\|f^{-1}\| = e^{|t'|}$, hence $\|f^{-1}\| = e^{d^{\mathrm{L}}(x_0, f(x_0))}$. From this remark and from the proof of (i), one obtains

$$\|\gamma_n y_0\|\|\gamma_n^{-1}\| = \|y_0\|e^{-B_{D_0}(x_0, \gamma_n^{-1}(x_0)) + d^{\mathrm{L}}(x_0, \gamma_n^{-1}(x_0))}.$$

It follows that the conditions

$$\lim_{n \to +\infty} \|\gamma_n^{-1}\| = +\infty \quad \text{and} \quad (\|\gamma_n y_0\|\|\gamma_n^{-1}\|)_{n \geqslant 1} \text{ is bounded}$$

are equivalent to the conditions

$$\lim_{n \to +\infty} d^L(x_0, \gamma_n^{-1}(x_0)) = +\infty \quad \text{and}$$

$$(-B_{D_0}(x_0, \gamma_n^{-1}(x_0)) + d^L(x_0, \gamma_n^{-1}(x_0)))_{n \geqslant 1} \text{ is bounded.}$$

The last two conditions are equivalent to the fact that the point D is conical (Proposition I.3.15).

The case in which D is arbitrary is treated like it was in the proof of part (i), replacing D with $f(D_0)$ and using the relation $(*)$. $\square$

2 Lorentzian interpretation of the dynamics of the geodesic flow

We write $\widetilde{g}_{\mathbb{R}}$ to denote the geodesic flow on $T^1\mathcal{H}_{-1}^+$. By definition, if $(x, \overrightarrow{v}) \in T^1\mathcal{H}_{-1}^+$ and $(v(t))_{t \in \mathbb{R}}$ is the arclength parametrization of the oriented geodesic associated with $(z, \overrightarrow{v})$ (see Sect. III.1), one has (Fig. VI.7)

$$\widetilde{g}_{t'}((z, \overrightarrow{v})) = \left(v(t'), \frac{dv}{dt}(t')\right).$$

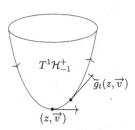

Fig. VI.7.

2.1 Lorentzian point of view on the set of trajectories of $\widetilde{g}_{\mathbb{R}}$

Let $(z, \overrightarrow{v})$ in $T^1\mathcal{H}_{-1}^+$ and $(v(t))_{t \in \mathbb{R}}$ be the oriented geodesic associated to $(z, \overrightarrow{v})$ such that $v(0) = z$. We denote by $D^-(v)$ and $D^+(v)$ respectively, the negative and positive endpoints of $(v(t))_{t \in \mathbb{R}}$.

Exercise 2.1. Prove that if $(z, \overrightarrow{v}) = ((0,0,1),(1,0,0))$, then $D^-(v) = \mathbb{R}(-1,0,1)$ and $D^+(v) = \mathbb{R}(1,0,1)$.

Let $u^-(v), u^+(v)$ be the direction vectors of $D^-(v)$ and $D^+(v)$ satisfying (Fig. VI.8)

$$\|u^-(v)\| = \|u^+(v)\| = 1 \quad \text{and} \quad u^-(v) \in \mathcal{H}_0^+, \quad u^+(v) \in \mathcal{H}_0^+.$$

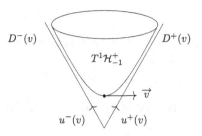

Fig. VI.8.

The vector $w(v) \in \mathcal{H}_1$ is defined as satisfying

$$b(w(v), w(v)) = 1, \quad b(w(v), u^-(v)) = 0, \quad b(w(v), u^+(v)) = 0 \quad \text{and}$$
$$(w(v), u^-(v), u^+(v)) \text{ is a positive basis.}$$

For example if $(z, \overrightarrow{v}) = ((0,0,1),(1,0,0))$, then

$$w(v) = (0,1,0).$$

Exercise 2.2. For all $f \in O^0(2,1)$, prove that

$$w(f(v)) = f(w(v)),$$

where $f(v)$ represents the geodesic $(f(v(t))_{t \in \mathbb{R}}$, and $f(w(v))$ represents the image of $w(v)$ by the linear map f.

Let W denote the map from $T^1\mathcal{H}_{-1}^+ \to \mathcal{H}_1$ defined by (Fig. VI.9):

$$W((z, \overrightarrow{v})) = w(v).$$

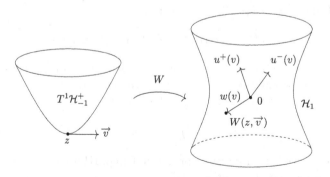

Fig. VI.9.

The map W satisfies the following properties.

Property 2.3.

(i) *The map* $W : T^1\mathcal{H}^+_{-1} \to \mathcal{H}_1$ *is continuous with respect to the metric* D^L *on* $T^1\mathcal{H}^+_{-1}$ *(see Exercise I.1.10).*

(ii) *The map* $W : T^1\mathcal{H}^+_{-1} \to \mathcal{H}_1$ *is surjective and*

$$W^{-1}(W((z, \overrightarrow{v}))) = \widetilde{g}_\mathbb{R}((z, \overrightarrow{v})).$$

Exercise 2.4. Prove Property 2.3.
(Hint: for (i), use the fact that the action of $O^0(2,1)$ is simply transitive on $T^1\mathcal{H}^+_{-1}$.)

Therefore, the map W induces a bijection between the set of trajectories of $\widetilde{g}_\mathbb{R}$ and the hyperboloid of one sheet $\mathcal{H}_1$.

In this model, the action of $O^0(2,1)$ on the set of geodesic trajectories corresponds to the linear action of this group on $\mathcal{H}_1$.

2.2 Linear action of a Lorentz group on $\mathcal{H}_1$ and the dynamics of the geodesic flow

Consider now a non-elementary Lorentz group Γ_L. Let Γ_F denote the Fuchsian group associated with Γ_L.

We use notations introduced in Chap. III. In our present context, π denotes the projection from $\mathcal{H}^+_{-1}$ to the surface $S = \Gamma_L \backslash \mathcal{H}^+_{-1}$, and π^1 denotes the projection from $T^1\mathcal{H}^+_{-1}$ to $T^1S = \Gamma_L \backslash T^1\mathcal{H}^+_{-1}$. Recall that the map P is the stereographic projection from $\mathcal{H}^+_{-1}$ to $\mathbb{D}$.

Take $f \in \Gamma_L$, and set $PfP^{-1} = \gamma$. For all $(z, \overrightarrow{v}) \in T^1\mathcal{H}^+_{-1}$, we have

$$f((z, \overrightarrow{v})) = \gamma((P(z), TP_z(\overrightarrow{v}))),$$

where TP_z is the tangent map of P at z. This map induces a homeomorphism φ from $T^1S \to \Gamma_F \backslash T^1\mathbb{D}$ defined by

$$\varphi(\Gamma_L((z, \overrightarrow{v}))) = \Gamma_F((P(z), TP_z(\overrightarrow{v}))).$$

The action of $O^0(2,1)$ on $T^1\mathcal{H}^+_{-1}$ commutes with that of $\widetilde{g}_\mathbb{R}$, thus this flow induces a flow, denoted $g_\mathbb{R}$, on T^1S, called the *geodesic flow*.

By construction, for all $\overline{(z, \overrightarrow{v})}$ in T^1S, one has

$$\varphi(g_t(\overline{(z, \overrightarrow{v})})) = g_t(\varphi(\overline{(z, \overrightarrow{v})})).$$

Our purpose now is to use results proved in Chap. III about the geodesic flow to obtain properties of the orbits of Γ_L on $\mathcal{H}_1$.

The following proposition, which is analogous to Proposition III.1.6, connects these two worlds.

Proposition 2.5. *Let u_1 and u_2 be in $\mathcal{H}_1$, and let $(z_1, \overrightarrow{v_1})$ and $(z_2, \overrightarrow{v_2})$ be elements of $T^1\mathcal{H}^+_{-1}$ such that $W((z_1, \overrightarrow{v_1})) = u_1$ and $W((z_2, \overrightarrow{v_2})) = u_2$. Consider a sequence $(\gamma_n)_{n\geqslant 1}$ in Γ_L. The following are equivalent:*

(i) *the sequence $(\gamma_n(u_1))_{n\geqslant 1}$ converges to u_2;*
(ii) *there exists a sequence $(s_n)_{n\geqslant 1}$ in $\mathbb{R}$ such that $(\widetilde{g}_{s_n}(\gamma_n((z_1, \overrightarrow{v_1}))))_{n\geqslant 1}$ converges to $(z_2, \overrightarrow{v_2})$.*

Exercise 2.6. Prove Proposition 2.5.
(Hint: use Exercises 2.2 and 2.3, and reuse the arguments from the proof of Proposition III.1.6.)

Let us now focus on the closed orbits of Γ_L in $\mathcal{H}_1$.

Exercise 2.7. Let $u \in \mathcal{H}_1$. Prove that $\Gamma_L(u)$ is closed if and only if every sequence in $\Gamma_L(u)$ converging in $\mathcal{H}_1$ is constant after some initial terms.

Let $u_1 \in \mathcal{H}_1$ and $(z_1, \overrightarrow{v_1}) \in W^{-1}(u_1)$. In Chap. III, we proved that the existence of a convergent sequence $(\widetilde{g}_{s_n}(\pi^1((z, \overrightarrow{v}))))_{n\geqslant 1}$, with $(s_n)_{n\geqslant 1}$ unbounded, is equivalent to the fact that $v(+\infty)$ or $v(-\infty)$ is conical (Proposition III.2.8). This result, added to Proposition 2.5 and Exercise 2.7, implies that if $\Gamma_L(u_1)$ is not closed then $D^-(v_1)$ or $D^+(v_1)$ is conical.

The converse is not true, because if $\pi^1((z_1, \overrightarrow{v_1}))$ is periodic, in other words if there exists $\gamma \in \Gamma_L - \{\mathrm{Id}\}$ fixing u_1, then the trajectory of $\pi^1((z_1, \overrightarrow{v_1}))$ is compact, and hence, by Proposition 2.5, the orbit $\Gamma_L(u_1)$ is closed.

However, if $\pi^1((z_1, \overrightarrow{v_1}))$ is not periodic and if either $D^-(v_1)$ or $D^+(v_1)$ is conical, then $\Gamma_L(u_1)$ is not closed. Indeed, under these hypotheses, there exist an unbounded sequence $(s_n)_{n\geqslant 1}$ and a sequence $(\gamma_n)_{n\geqslant 1}$ in Γ_L such that $(\widetilde{g}_{s_n}(\gamma_n((z_1, \overrightarrow{v_1}))))_{n\geqslant 1}$ converges, and such that the sequence of trajectories $(\widetilde{g}_{\mathbb{R}}(\gamma_n((z_1, \overrightarrow{v_1}))))_{n\geqslant 1}$ is not stationary. This implies, by Proposition 2.5 and Exercise 2.7, that $\Gamma_L(u_1)$ is not closed.

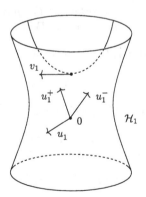

Fig. VI.10.

Let u_1^+ (respectively u_1^-) denote the element of $\mathcal{H}_0^+$ of Euclidean norm 1 satisfying (Fig. VI.10)

$$D^+(v_1) = \mathbb{R}u_1^+ \quad (\text{respectively } D^-(v_1) = \mathbb{R}u_1^-).$$

The preceding argument, together with the Lorentzian characterization of conical points (Proposition 1.7), implies the following result:

Theorem 2.8. *Let $u_1 \in \mathcal{H}_1$. The orbit $\Gamma_{\mathrm{L}}(u_1)$ is closed if and only if one of the following conditions is satisfied:*

(i) *there exists γ in $\Gamma_{\mathrm{L}} - \{\mathrm{Id}\}$ such that $\gamma(u_1) = u_1$;*
(ii) *for every sequence $(\gamma_n)_{n \geqslant 1}$ in Γ_{L} satisfying $\lim_{n \to +\infty} \|\gamma_n^{-1}\| = +\infty$, the sequences $(\|\gamma_n^{-1}\| \|\gamma_n(u_1^-)\|)_{n \geqslant 1}$ and $(\|\gamma_n^{-1}\| \|\gamma_n(u_1^+)\|)_{n \geqslant 1}$ are not bounded.*

One can deduce from this theorem and from Corollary I.4.17 the following characterization of Lorentz lattices in terms of their action on $\mathcal{H}_1$.

Corollary 2.9. *A Lorentz group Γ_{L} is a lattice if and only if the only closed orbits of Γ_{L} in $\mathcal{H}_1$ are either orbits of vectors which are fixed by a hyperbolic isometry of Γ_{L}, or are orbits of the form $\Gamma_{\mathrm{L}}(u)$, where $\mathbb{R}u^-$ and $\mathbb{R}u^+$ are eigenlines of a parabolic isometry in Γ_{L}.*

Note that, if $L(\Gamma_{\mathrm{L}}) \neq \mathcal{H}_{-1}^+(\infty)$, which is the case for example if Γ_{F} is a Schottky group (see Chap. II) then, by Proposition 1.7 and Theorem 2.8, if $u \in \mathcal{H}_1$ and if $\mathbb{R}u^-$ and $\mathbb{R}u^+$ do not belong to $L(\Gamma_{\mathrm{L}})$, then the orbit of u under Γ_{L} is closed.

A Lorentz group is *geometrically finite* if $L(\Gamma_{\mathrm{L}})$ consists entirely of conical or parabolic points (see Theorem I.4.13). Corollary 2.9 can be generalized to this family of groups in the following way.

Corollary 2.10. *A Lorentz group Γ_{L} is geometrically finite if and only if the only closed orbits of Γ_{L} on $\mathcal{H}_1$ are either the orbits of vectors fixed by an hyperbolic isometry of Γ_{L}, or are orbits of the form $\Gamma_{\mathrm{L}}(u)$, where $\mathbb{R}u^-$ and $\mathbb{R}u^+$ are in the set $L_p(\Gamma_{\mathrm{L}}) \cup (\mathcal{H}_{-1}^+(\infty) - L(\Gamma_{\mathrm{L}}))$.*

Let us introduce the set $\mathcal{H}_1(\Gamma_{\mathrm{L}})$ defined by

$$\mathcal{H}_1(\Gamma_{\mathrm{L}}) = \{u \in \mathcal{H}_1 \mid \mathbb{R}u^- \text{ and } \mathbb{R}u^+ \text{ are in } L(\Gamma_{\mathrm{L}})\}.$$

Exercise 2.11. Prove that $\mathcal{H}_1(\Gamma_{\mathrm{L}})$ is closed and invariant with respect to Γ_{L}.

The set $\mathcal{H}_1(\Gamma_{\mathrm{L}})$ is related to the non-wandering set of the geodesic flow (Theorem III.2.1) in the following way:

$$\pi^1(W^{-1}(\mathcal{H}_1(\Gamma))) = \Omega_g(T^1 S).$$

As we have shown in Chaps. III and IV, the topological nature of a non-closed trajectory of the geodesic flow on $\Omega_g(T^1 S)$ may be very complex. It follows that, if $u \in \mathcal{H}_1(\Gamma)$ and $\Gamma_L(u)$ is not closed, no information about the closure of $\Gamma_L(u)$ is available without additional hypotheses on u.

However, one can state the following properties which follow directly from Theorems III.3.3, III.4.2 and from the continuity of the map W:

Property 2.12. *Let Γ_L be a non-elementary Lorentz group.*

(i) *The set of vectors in $\mathcal{H}_1$ fixed by hyperbolic isometries of Γ_L is dense in $\mathcal{H}_1(\Gamma_L)$.*

(ii) *Some orbits of Γ_L are dense in $\mathcal{H}_1(\Gamma_L)$.*

3 Lorentzian interpretation of the dynamics of the horocycle flow

We continue to use notations introduced in the preceding section. In Chap. V we established a correspondence between the set of trajectories of the horocycle flow on $T^1\mathbb{H}$ and $\{\pm \mathrm{Id}\}\backslash\mathbb{R}^2 - \{0\}$. The Lorentzian model that we propose below brings the methods that were used in Chap. V into play. For this reason, many of the proofs are left as exercises.

We will again write $\tilde{h}_\mathbb{R}$ to denote the horocycle flow on $T^1\mathcal{H}_{-1}^+$ defined by

$$\tilde{h}_{t'}((z, \overrightarrow{v})) = (\beta(t'), \overrightarrow{v'}),$$

where $(\beta(t))_{t\in\mathbb{R}}$ is the arclength parametrization of the horocycle centered at $D^+(v)$ passing through z for which one has $\beta(0) = z$, and for which the ordered pair $(d\beta/dt(t'), \overrightarrow{v'})$ is a positive orthonormal basis for $T_z\mathcal{H}_{-1}^+$ (Fig. VI.11).

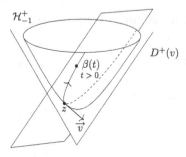

Fig. VI.11.

3.1 Lorentzian point of view on the set of trajectories of $\tilde{h}_\mathbb{R}$

Consider the map V, from $T^1\mathcal{H}_{-1}^+$ into the positive half cone $\mathcal{H}_0^{+*}$, defined by

$$V((z, \overrightarrow{v})) = e^{B_{D^+(v)}(x_0,z)/2}u^+(v),$$

where $x_0 = (0,0,1)$ and $u^+(v)$ is the unit vector (in the Euclidean norm) belonging to $\mathcal{H}_0^+$ and $D^+(v)$.

For example, if $z = (0,0,1)$ and $\overrightarrow{v} = (1,0,0)$, then $V((z,\overrightarrow{v})) = (1/\sqrt{2})(1,0,1)$ (Fig. VI.12).

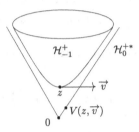

Fig. VI.12.

Property 3.1. *Let $f \in O^0(2,1)$ and $(z,\overrightarrow{v}) \in T\mathcal{H}_{-1}^+$. The following properties are satisfied*

(i) $V(f((z,\overrightarrow{v}))) = f(V((z,\overrightarrow{v})))$;
(ii) *the map* $V : T^1\mathcal{H}_{-1}^+ \to \mathcal{H}_0^{+*}$ *is surjective and* $V^{-1}(V((z,\overrightarrow{v}))) = \widetilde{h}_{\mathbb{R}}((z,\overrightarrow{v}))$;
(iii) *the map* $V : T^1\mathcal{H}_{-1}^+ \to \mathcal{H}_0^{+*}$ *is continuous.*

Exercise 3.2. Prove Property 3.1.
(Hint: see Exercises V.1.6, V.1.8 and Proposition V.1.7.)

3.2 Linear action of a Lorentz group on $\mathcal{H}_0^{+*}$ and dynamics of the horocycle flow

As in Sect. 2.2 of this chapter, we consider a non-elementary Lorentz group Γ_{L}. Since the action of $O^0(2,1)$ commutes with that of $\widetilde{h}_{\mathbb{R}}$, this flow induces another flow, denoted $h_{\mathbb{R}}$, on $T^1S = \Gamma_{\mathrm{L}} \backslash T^1\mathcal{H}_{-1}^+$. The following proposition allows us to establish a relationship between the topological behavior of the orbits of Γ_{L} on $\mathcal{H}_0^{+*}$ and that of the trajectories of $h_{\mathbb{R}}$.

Proposition 3.3. *Let $u_1, u_2 \in \mathcal{H}_0^{+*}$ and $(z_1,\overrightarrow{v_1}), (z_2,\overrightarrow{v_2}) \in T^1\mathcal{H}_{-1}^+$ such that $D^+(v_i) = \mathbb{R}u_i$ for $i = 1,2$. Consider a sequence $(\gamma_n)_{n\geqslant 1}$ in Γ_{L}. The following are equivalent:*

(i) *the sequence $(\gamma_n(u_1))_{n\geqslant 1}$ converges to u_2;*
(ii) *there exists a sequence $(s_n)_{n\geqslant 1}$ in $\mathbb{R}$ such that $\lim_{n\to+\infty} \widetilde{h}_{s_n}(\gamma_n((z_1,\overrightarrow{v_1}))) = (z_2,\overrightarrow{v_2})$.*

Exercise 3.4. Prove Proposition 3.3.
(Hint: Rewrite the proof of Proposition V.2.1.)

Consider the set $\mathcal{H}_0^+(\Gamma_L)$ defined as follow

$$\mathcal{H}_0^+(\Gamma_L) = \{u \in \mathcal{H}_0^{+*} \mid \mathbb{R}u \in L(\Gamma_L)\}.$$

Exercise 3.5. Prove that $\mathcal{H}_0^+(\Gamma_L)$ is closed in $\mathcal{H}_0^{+*}$ and invariant with respect to Γ_L.

This set is related to the non-wandering set of the horocycle flow (Proposition V.2.11) by the following equality:

$$\pi^1(V^{-1}(\mathcal{H}_0^+(\Gamma_L))) = \Omega_h(T^1S).$$

In particular, $\mathcal{H}_0(\Gamma_L)$ contains the isotropic eigenvectors of the parabolic and hyperbolic isometries of Γ_L.

Exercise 3.6. Prove that, if $u \in \mathcal{H}_0^{+*}$ is fixed by a parabolic isometry of Γ_L or if u does not belong to $\mathcal{H}_0^+(\Gamma_L)$, then $\Gamma_L(u)$ is closed in $\mathcal{H}_0^{+*}$.

In Chap. V, we proved that the trajectory of $\pi^1((z, \vec{v}))$ is dense in $\Omega_h(T^1S)$ if and only if $v(+\infty)$ is horocyclic. This result, together with Proposition 1.7 and Lemma 3.3, allows us to state the following proposition.

Proposition 3.7. *Let $u \in \mathcal{H}_0^{+*}$. The Γ_L-orbit of u is dense in $\mathcal{H}_0(\Gamma_L)$ if and only if there exists $(\gamma_n)_{n \geqslant 1}$ in Γ_L such that $\lim_{n \to +\infty} \gamma_n(u) = 0$.*

Exercise 3.8. Let $u \in \mathcal{H}_0^{+*}$.

(i) Prove that if $\Gamma_L(u)$ is closed in $\mathcal{H}_0^{+*}$, then $\Gamma_L(u)$ is closed in $\mathcal{H}_0$.
(ii) Prove that $\Gamma_L(u)$ is closed in $\mathcal{H}_0^{+*}$ if and only if every sequence $\Gamma_L(u)$ converging in $\mathcal{H}_0$ is stationary.

If Γ_L is a lattice, then $\mathcal{H}_0(\Gamma_L) = \mathcal{H}_0^{+*}$. Moreover, if Γ_L is uniform, then $\mathcal{H}_{-1}^+(\infty)$ consists entirely of horocyclic points, and Proposition 3.7 implies that every orbit of Γ_L on $\mathcal{H}_0^+$ is dense.

More generally, when Γ_L is geometrically finite, Proposition V.4.3 translated into the Lorentzian context, becomes the following proposition.

Proposition 3.9.

(i) *The group Γ_L is geometrically finite and non-elementary if and only if for all $u \in \mathcal{H}_0(\Gamma_L)$, either $\Gamma_L(u)$ is dense in $\mathcal{H}_0^+(\Gamma_L)$ or u is fixed by a parabolic isometry of Γ_L.*
(ii) *The group Γ_L is a uniform lattice if and only if every orbit of Γ_L on $\mathcal{H}_0^+$ is dense in $\mathcal{H}_0^+$.*

4 Comments

Following the work of G. Hedlund [37] and L. Greenberg [4], the study of orbits of groups acting linearly on a vector space has become a research area in its own right. One of the first results in this area was the equivalence, for lattices in $SL(n, \mathbb{R})$, between the presence of the zero vector in the closure of a non-trivial orbit and the density of that orbit in $\mathbb{R}^n$ [4]. This result has since been extended by J.-P. Conze and Y. Guivarc'h to some discrete subgroups of $SL(n, \mathbb{R})$ [16]. For $n = 2$, it has also been proved in Chap. V.

In this chapter (and Sect. V.2), we have introduced a relationship between the linear action of a discrete subgroup Γ_L of $SO^0(2, 1)$ on $\mathbb{R}^3$ and the dynamics of the geodesic or horocycle flow on $\Gamma_L \backslash T^1 \mathcal{H}_{-1}^+$ (or $\Gamma_F \backslash T^1 \mathbb{H}$). This relationship is based on a change in point of view which consists of interpreting the linear action of Γ_L as an action on the set of trajectories of a flow on $T^1 \mathcal{H}_{-1}^+$ (or $T^1 \mathbb{H}$). We have also obtained a correspondence between the topology of linear orbits and that of the trajectories of these flows, which has been used for example to prove the existence of dense trajectories in the non-wandering set of the horocycle flow (Corollary V.2.10).

In the metric context, there are many applications of this new point of view ([5, 32] and [60, Chap. II]). One example is the study of the asymptotic behavior of the number of vectors in $\mathcal{H}_{-1} \cap \mathbb{Z}^3$ having Euclidean norm $\leqslant T$, discussed in M. Babillot's text [5, Sect. 3.2], which reduces to counting the points of an orbit of the group $SL(3, \mathbb{Z}) \cap SO^0(2, 1)$ in a disk on the surface $\mathcal{H}_{-1}^+$ equipped with the metric g^L.

VII

Trajectories and Diophantine approximations

In this chapter, our setting is the Poincaré half-plane. Consider a non-elementary, geometrically finite Fuchsian group Γ (see Chap. I for the definitions) which *contains a non-trivial translation*. With these hypotheses, the surface $S = \Gamma \backslash \mathbb{H}$ admits finitely many cusps (see Sects. I.3 and I.4) (Fig. VII.1).

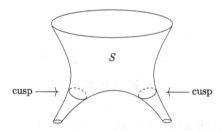

Fig. VII.1.

As in the previous chapters, we let π denote the projection from $\mathbb{H}$ to S. In the first step, we study the excursions of a geodesic ray $\pi([z, x))$ into the cusp corresponding to the image of the restriction of π to a horodisk centered at the point ∞. Our purpose is to relate the frequency of these excursions to the way in which the real number x is approximated by the Γ-orbit of the point ∞.

In the second step, we restrict our attention to the modular group and rediscover, in the spirit of Chap. III, some classical results of the theory of Diophantine approximations.

F. Dal'Bo, *Geodesic and Horocyclic Trajectories*, Universitext,
DOI 10.1007/978-0-85729-073-1_7, © Springer-Verlag London Limited 2011

1 Excursions of a geodesic ray into a cusp

Let us begin by noting that the point ∞ is a parabolic point of the limit set of Γ since this group contains a translation.

For all $t > 0$, set

$$H_t = \{z \in \mathbb{H} \mid \operatorname{Im} z = t\} \quad \text{and} \quad H_t^+ = \{z \in \mathbb{H} \mid \operatorname{Im} z \geqslant t\}.$$

The set H_t is the horocycle centered at ∞, which is the level curve for the value $\ln t$ of the function $f(z) = B_\infty(i, z)$. The set H_t^+ is its corresponding horodisk.

Recall that there exists $t_0 > 0$ such that for all γ in $\Gamma - \Gamma_\infty$, one has

$$\gamma H_{t_0} \cap H_{t_0} = \varnothing \quad \text{(Theorem I.3.17)}.$$

With this condition, the projection from $\Gamma_\infty \backslash H_{t_0}^+$ to the cusp $\pi(H_{t_0}^+)$ is injective (see the end of Sect. I.3) (Fig. VII.2).

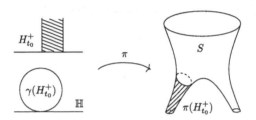

Fig. VII.2.

Let us now fix such a horodisk $H_{t_0}^+$ and a point $z \in \mathbb{H}$. Take $x \in \mathbb{H}(\infty)$ and denote $(r(s))_{s \geqslant 0}$ the arclength parametrization of the ray $[z, x)$.

Does the geodesic ray $\pi([z, x))$ visit the cusp $\pi(H_{t_0}^+)$? As we will see, the answer depends on properties of the point x.

Suppose first that, for some $T \geqslant 0$, the ray $\pi([r(T), x))$ is in $\pi(H_{t_0}^+)$. This implies that the ray $[r(T), x)$ is covered by the family $\gamma(H_{t_0}^+)$ with $\gamma \in \Gamma - \Gamma_\infty$. Since these horodisks are disjoint, this ray is contained in only one of these horodisks, which implies that $x = \gamma(\infty)$ for some $\gamma \in \Gamma$.

Conversely if $x = \gamma(\infty)$, then there exists $T \geqslant 0$ such that $\pi([r(T), x))$ is in $\pi(H_{t_0}^+)$, by Theorem III.2.11.

Thus one obtains:

Proposition 1.1. *There exists $T \geqslant 0$ such that $\pi([r(T), x))$ is contained in $\pi(H_{t_0}^+)$ if and only if x belongs to the Γ-orbit of the point ∞.*

Furthermore, by Theorem III.2.11, if $\pi([r(T), x))$ is in $\pi(H_{t_0}^+)$, then the map from $[T, \infty)$ to $\pi(H_{t_0}^+)$ which sends a real number s to $\pi(r(s))$ is an embedding (Fig. VII.3).

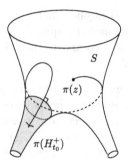

Fig. VII.3. $x \in \Gamma(\infty)$

Suppose now that the point x is not in $\Gamma(\infty)$. According to Proposition 1.1, the ray $\pi([z, x))$ is not contained in the horodisk $\pi(H_{t_0}^+)$. As consequence, we obtain

Corollary 1.2. *Suppose that $x \in \mathbb{H}(\infty)$ does not belong to $\Gamma(\infty)$. Let $t > 0$ be such that $\gamma(H_t) \cap H_t = \varnothing$ for all $\gamma \in \Gamma - \Gamma_\infty$. If there exists $s \geqslant 0$ such that $\pi(r(s)) \in \pi(H_t^+)$, then there exists $s' > s$ such that $\pi(r(s')) \in \pi(H_t)$ (Fig. VII.4).*

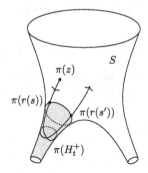

Fig. VII.4. $x \notin \Gamma(\infty)$ and $\pi(r(s)) \in \pi(H_t^+)$

Corollary 1.3. *Let $t_0 > 0$, if x does not belong to $\Gamma(\infty)$ and is not conical, then there exists $T \geqslant 0$ such that $\pi([r(T), x)) \cap \pi(H_{t_0}^+) = \varnothing$.*

Proof. Since x does not belong to $\Gamma(\infty)$, then, by Proposition 1.1, two cases arise:

(i) either there exists an unbounded sequence $(s_n)_{n \geqslant 1}$ such that $\pi(r(s_n)) \in \pi(H_{t_0}^+)$;

(ii) or there exists $T \geqslant 0$ such that $\pi([r(T), x)) \cap \pi(H_{t_0}^+) = \varnothing$.

Let us prove that the case (i) implies that x is conical. Actually, by Corollary 1.2, there exists an unbounded sequence $(s'_n)_{n\geqslant 1}$ such that $\pi(r(s'_n)) \in \pi(H_{t_0})$. Since the set $\pi(H_{t_0})$ is compact, then, by Proposition III.2.8 (on S instead of $\Gamma\backslash T^1\mathbb{H}$), x is conical.   □

Let us analyze the topology of a ray $\pi([z,x))$. Since the group Γ is geometrically finite, if x is not conical then either x is parabolic or x does not belong to $L(\Gamma)$. In the first case, by the preceding argument, there exists $T \geqslant 0$ such that the map from $[T,\infty)$ into a cusp $\pi(H_t^+(x))$ associated with x, which maps the real number s to $\pi(r(s))$, is an embedding (Fig. VII.5).

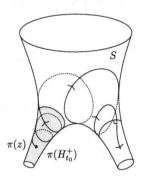

Fig. VII.5. $x \notin \Gamma(\infty)$ and x is parabolic

The second case is the subject of the following exercise.

Exercise 1.4. Prove that if x does not belong to $L(\Gamma)$, then there exists $T \geqslant 0$ such that the map from $[T,+\infty)$ into S which sends a real number s to $\pi(r(s))$ is an embedding (Fig. VII.6).
(Hint: see Property III.2.10.)

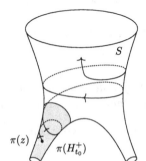

Fig. VII.6. $x \notin L(\Gamma)$

It now remains to address the case where x is conical. Note that if x is the fixed point of a hyperbolic isometry in Γ and if z is on the axis of that isometry, then $\pi([z, x))$ is a compact geodesic in S and thus, for sufficiently large t_0, this ray does not intersect the cusp $\pi(H_{t_0}^+)$ (Fig. VII.7).

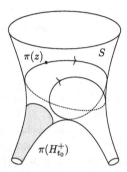

Fig. VII.7. $\pi([z, x))$ compact geodesic

On the other hand, we will see for example when Γ is the modular group, that there exist irrational numbers x such that $\pi([z, x))$ is not bounded. In this case, there exists an unbounded sequence $(s_n)_{n \geqslant 0}$ for which $\pi(r(s_n))$ belongs to $\pi(H_{t_0}^+)$ (Fig. VII.8).

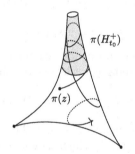

Fig. VII.8. $\Gamma = \mathrm{PSL}(2, \mathbb{Z})$

Thus the fact that $x \in \mathbb{H}(\infty)$ is conical cannot be characterized solely in terms of excursions of the ray $\pi([z, x))$ into $\pi(H_{t_0}^+)$. It is necessary to consider the family of horodisks $\pi(H_t^+)$ with $t \in \mathbb{R}_+^*$.

We associate to each $x \in \mathbb{H}$ the set $E([z, x)) \subset \mathbb{R}^{*+}$, of $t > 0$ for which there exists an unbounded sequence $(s_n)_{n \geqslant 1}$ satisfying

$$\pi(r(s_n)) \in \pi(H_t).$$

If $x \in \Gamma(\infty)$ or if $x \notin \Gamma(\infty)$ and is not conical, then $E([z, x)) = \varnothing$ (Proposition 1.3). This is not the case if x is conical.

Proposition 1.5.

(i) *There exists* $t_1 > 0$ *such that for every conical point* x *in* $L(\Gamma)$, *one has* $t_1 \in E([z, x))$.

(ii) *If* x *is a conical point in* $L(\Gamma)$, *the upper bound in* $\mathbb{R}^+ \cup \{+\infty\}$ *of the set* $E([z, x))$ *is independent of* z.

Proof.

(i) We suppose that x is conical. The group Γ is geometrically finite and non-elementary, therefore by Proposition III.2.11 on S (rather than $T^1 S$), there exists a compact subset $K_1 \subset S$ (independent of x) and an unbounded sequence $(s_n)_{n \geqslant 1}$ such that $\pi(r(s_n))$ belongs to K_1. Let us lift K_1 to a compact set $\widetilde{K}_1$ included in a horodisk $H_{t_1}^+$. By Property 1.2, there exists $s_n' \geqslant s_n$ satisfying $\pi(r(s_n')) \in \pi(H_{t_1})$. This shows that t_1 belongs to $E([z, x))$.

(ii) Let x be a conical point in $L(\Gamma)$. Take $z' \neq z$ in $\mathbb{H}$, and denote by $(r'(s))_{s \geqslant 0}$ the arclength parametrization of the ray $[z', x)$. Notice that for all $\varepsilon > 0$, there exists $T \geqslant 0$ such that, $[r'(T), x)$ is in the ε-neighborhood of the ray $[z, x)$.

Fix a real number t in $E([z, x))$. There exists a sequence $(\gamma_n)_{n \geqslant 1}$ in Γ and an unbounded sequence $(s_n)_{n \geqslant 1}$ such that $\gamma_n(r(s_n))$ belongs to H_t, in other words

$$B_\infty(i, \gamma_n(r(s_n))) = t.$$

Let $\varepsilon > 0$ and let $(s_n')_{n \geqslant 1}$ be a sequence satisfying $d(r'(s_n'), r(s_n)) \leqslant \varepsilon$. Using Property I.1.19 of the Busemann cocycle, one obtains the following statement:

$$t - \varepsilon \leqslant B_\infty(i, \gamma_n(r'(s_n))) \leqslant t + \varepsilon.$$

It follows that $\pi(r'(s_n))$ belongs to $\pi(H_{t-\varepsilon}^+)$. Thus there exists $s_n'' \geqslant s_n'$ such that $\pi(r'(s_n''))$ belongs to $\pi(H_{t-\varepsilon})$, which shows that $t-\varepsilon$ belongs to $E([z', x))$. In conclusion, the upper bound of this set is at least that of $E([z, x))$. Changing the roles of z and z' completes the proof. □

Definition 1.6. Let $x \in L(\Gamma)$ be a conical point. The upper bound of the set $E([z, x))$ is called the *height* of the ray $\pi([z, x))$ and is written $h(x)$. Moreover, x is said to be *geometrically badly approximated* if its height $h(x)$ is finite.

If $\pi([z, x))$ is bounded, clearly x is geometrically badly approximated. Is the converse true? In the following section, we give an answer to this question.

2 Geometrically badly approximated points

Consider a conical point $x \in L(\Gamma)$ which is geometrically badly approximated, and a real number $t > h(x)$ satisfying

$$\gamma(H_t^+) \cap H_t^+ = \varnothing \quad \text{for all } \gamma \in \Gamma - \Gamma_\infty.$$

By definition of $h(x)$, there exists $T > 0$ such that $\pi([r(T), x))$ does not intersect $\pi(H_t)$. For t large enough, the horocycle $\pi(H_t)$ separates S into two connected components, therefore two cases arise:

(i) $\pi([r(T), x)) \subset \pi(H_t^+)$;
(ii) $\pi([r(T), x)) \subset S - \pi(H_t^+)$.

The first case is excluded by Property 1.2. It remains to consider case (ii). Since the set $\pi([z, r(T)])$ is compact, there exists $t' \geqslant t$ such that

$$\pi([z, x)) \cap \pi(H_{t'}^+) = \varnothing.$$

Thus one obtains the following characterization:

Proposition 2.1. *A conical point x in $L(\Gamma)$ is geometrically badly approximated if and only if there exists $t > 0$ such that*

$$\pi([z, x)) \cap \pi(H_t^+) = \varnothing.$$

Notice that this proposition does not imply that if x is geometrically badly approximated, then $\pi([z, x))$ is bounded. Actually, in the case where $L(\Gamma)$ contains a parabolic point y which is not in $\Gamma(\infty)$, then the surface S admits at least two disjoint cusps (see for example the group $\Gamma(2)$ from Chap. II), thus the ray $\pi([z, x))$ can be unbounded without meeting a cusp $C(H_t^+)$ (Fig. VII.9).

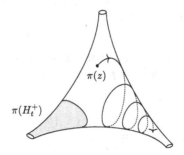

Fig. VII.9. $\Gamma = \Gamma(2)$

On the other hand, if the set of parabolic points of $L(\Gamma)$ is reduced to the Γ-orbit of the point ∞, then the projection to S of the Nielsen region $N(\Gamma)$ is the union of a compact set and $\pi(H_t^+)$ (for large t) (Proposition I.4.16). Therefore, if x is geometrically badly approximated, then $\pi([z, x))$ is bounded.

Corollary 2.2. *If the set of parabolic points of $L(\Gamma)$ is equal to $\Gamma(\infty)$, then a conical point x in $L(\Gamma)$ is geometrically badly approximated if and only if the ray $\pi([z, x))$ is bounded.*

Now consider the particular case where the group Γ is a Schottky group $S(p, h)$ generated by a translation p and a hyperbolic isometry h (see Sect. II.1) (Fig. VII.10).

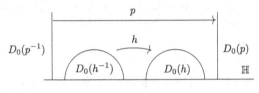

Fig. VII.10. $\Gamma = S(p, h)$

As we saw in Chap. II (Property II.1.9), under these hypotheses, the set of parabolic points of $S(p, h)$ is equal to the orbit of the point ∞, and hence Corollary 2.2 applies. Recall from Proposition II.2.2 that a conical point of $L(S(p, h))$ is uniquely represented by a sequence $f(x) = (s_i)_{i \geqslant 1}$ satisfying

$$s_i \in \{h^{\pm 1}, p^{\pm 1}\}, \quad s_{i+1} \neq s_i^{-1}$$

and if $s_i \in \{p^{\pm 1}\}$, then there exists $j > i$ such that $s_j \in \{h^{\pm 1}\}$.

The following proposition characterizes the geometrically badly approximated points in coding terms.

Proposition 2.3. *Let x be a conical point in $L(S(p, h))$. Define $f(x) = (s_i)_{i \geqslant 1}$. The following statements are equivalent:*

(i) *the point x is geometrically badly approximated;*
(ii) *there exists an integer $r > 0$ such that if $s_i \in \{p^{\pm 1}\}$, then there exists $1 \leqslant j \leqslant r$ such that $s_{i+j} \in \{h^{\pm 1}\}$.*

Proof.
Not (ii) $\Rightarrow$ not (i). Consider the sequence $(a_i)_{i \geqslant 1}$ constructed from $f(x)$ by grouping together the consecutive p and p^{-1} terms. Such a sequence satisfies

$$a_i \in \{p^n, h^{\pm 1} \mid n \in \mathbb{Z}^*\}, \quad a_i \neq a_{i+1}^{-1},$$
$$\text{if } a_i = p^n \text{ then } a_{i+1} \in \{h^{\pm 1}\},$$
$$\text{and} \quad x = \lim_{n \to +\infty} a_1 \cdots a_n(z_0),$$

where z_0 is a point in $\mathbb{H} - (D_0(h) \cup D_0(h^{-1}) \cup D_0(p) \cup D_0(p^{-1}))$. By hypothesis, there exists a subsequence $(a_{i_k})_{k \geqslant 1}$ such that $a_{i_k} = p^{n_k}$ with $\lim_{k \to +\infty} |n_k| = +\infty$. Define $\gamma_k = a_1 \cdots a_{i_k}$. We have,

$$\lim_{k \to +\infty} \gamma_k^{-1}(z_0) = \infty \quad \text{and} \quad \gamma_k^{-1}(x) \in \overline{D_0(h) \cup D_0(h^{-1})}$$

(see Property II.1.4).

The points $\gamma_k^{-1}(x)$ belong to a compact subset of $\mathbb{R}$ hence, passing to a subsequence, one can assume that the sequence of rays $[\gamma_k^{-1}(z_0), \gamma_k^{-1}(x))$ converges to a geodesic (∞y). It follows that for all $t > 0$, there exists $k \geqslant 1$ such that

$$[\gamma_k^{-1}(i), \gamma_k^{-1}(x)) \cap H_t^+ \neq \varnothing.$$

This property implies that the ray $\pi([z_0, x))$ is not bounded, and therefore that x is not geometrically badly approximated.

Not (i) $\Rightarrow$ not (ii). Suppose that x is not geometrically badly approximated and choose z from the geodesic (∞x). By assumption, there exists a sequence $(t_n)_{n \geqslant 1}$ which converges to $+\infty$ and a sequence $(\gamma_n)_{n \geqslant 1}$ in $S(p, h)$ such that

$$\gamma_n([z, x)) \cap H_{t_n} \neq \varnothing.$$

Passing to a subsequence, one can assume that γ_n can be written in the form of a reduced word $c_1 \cdots c_{\ell_n}$ satisfying

$$c_1 \in \{h^{\pm 1}\}, \quad c_i \in \{p^{\pm 1}, h^{\pm 1}\} \quad \text{for } i \geqslant 2, \quad \text{and} \quad \lim_{n \to +\infty} \ell_n = +\infty.$$

The point $\gamma_n(\infty)$ belongs to $\overline{D_0(h) \cup D_0(h^{-1})} \cap \mathbb{R}$, which is a compact subset of $\mathbb{R}$, and the sequence of radii of Euclidean circular arcs $(\gamma_n(\infty)\gamma_n(x))$ converges to $+\infty$ thus

$$\lim_{n \to +\infty} \gamma_n(x) = \infty.$$

This property implies the existence of $N_1 > 0$ such that

$$\forall n \geqslant N_1, \quad c_1 = s_{\ell_n}^{-1}, \ldots, c_n = s_1^{-1}.$$

To see this, note that $x = \lim_{n \to +\infty} s_1 \cdots s_n(z_0)$. If the preceding condition is not satisfied, there exists a subsequence $(\gamma_{n_p})_{p \geqslant 1}$ such that the first letter of the reduced word corresponding to $\gamma_{n_p} s_1 \cdots s_{\ell_{n_p}}$ is the letter c_1. In this case, by Property II.1.4(i), $\gamma_{n_p}(x)$ belongs to the compact set $\overline{D_0(h) \cup D_0(h^{-1})} \cap \mathbb{R}$, which is not allowed.

It follows that for $n \geqslant N_1$, the point $\gamma_n(x)$ belongs to $D(s_{\ell_n+1})(\infty)$. Since $\lim_{n \to +\infty} \gamma_n(x) = \infty$, there exists $N_2 \geqslant N_1$ such that

$$\forall n \geqslant N_2, \quad s_{\ell_n+1} \in \{p^{\pm 1}\}.$$

Let $n \geqslant N_2$. The point $s_{\ell_n+1}^{-1} \gamma_n(x)$ belongs to $D(s_{\ell_n+2})(\infty)$ and one has $\lim_{n \to +\infty} s_{\ell_n+1}^{-1} \gamma_n(x) = \infty$ thus, after reusing the same argument, there exists $N_3 \geqslant N_2$ such that

$$\forall n \geqslant N_3, \quad s_{\ell_n+1} \in \{p^{\pm 1}\} \quad \text{and} \quad s_{\ell_n+2} = s_{\ell_n+1}.$$

Continuing to apply this argument, one obtains an increasing sequence $(N_k)_{k \geqslant 2}$ satisfying the following condition:

$$\forall n \geqslant N_k, \quad s_{\ell_n+1} \in \{p^{\pm 1}\}, \quad s_{\ell_n+1} = \cdots = s_{\ell_n+k-1},$$

which contradicts part (i). $\qquad\square$

3 Applications to the theory of Diophantine approximations

We consider now the case where Γ is the modular group $\mathrm{PSL}(2,\mathbb{Z})$. The set of parabolic points associated with this group is simply the orbit of the point ∞ and is equal to $\mathbb{Q} \cup \{\infty\}$ (Property II.3.8). The modular surface $S = \mathrm{PSL}(2,\mathbb{Z})\backslash\mathbb{H}$ therefore admits a single type of cusp $\pi(H_t^+)$, where H_t^+ is a horodisk centered at the point ∞ (Fig. VII.11).

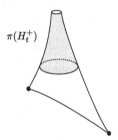

$\pi(H_t^+)$

Fig. VII.11. $\Gamma = \mathrm{PSL}(2,\mathbb{Z})$

The purpose of this section is to draw a parallel between the excursions into $\pi(H_t^+)$ of a geodesic ray $\pi([z, x))$, where x is irrational, and an approximation of x by a sequence of rational numbers.

We begin with a discussion of three well-known results from number theory. We will prove them in this section using a hyperbolic point of view.

3.1 Three classical theorems

The idea of continued fractions emerged very early [25, Chap. V]. In Sect. II.4, we gave a geometric interpretation of this idea using the Farey tiling of $\mathbb{H}$. One branch of the theory of Diophantine approximations consists of constructing a one-to-one correspondence between some algebraic properties of an irrational number and those of the sequence of integers $(n_i)_{i\geqslant 0}$ associated with its continued fraction expansion. One example is Proposition II.4.11 which relates the quadratic real numbers to almost periodic sequences.

Another branch focuses on the speed of convergence of the sequence of rational numbers associated with a continued fraction expansion. For example, one of these problems is to find the "best" (in the sense of asymptotic behavior) function $\psi : \mathbb{N} \to \mathbb{R}_+^*$ decreasing to 0 such that for all $x \in \mathbb{R} - \mathbb{Q}$, there exists a sequence of rational numbers $(p_n/q_n)_{n\geqslant 1}$ satisfying

$$|x - p_n/q_n| \leqslant \psi(|q_n|) \quad \text{and} \quad \lim_{n\to+\infty} |q_n| = +\infty.$$

Note that if p_n is the integer part of nx, the sequence $(p_n/n)_{n\geqslant 1}$ satisfies

$$|x - p_n/n| \leqslant 1/n.$$

It follow that the function $\psi(n)$ that we are looking for is thus less than $1/n$.

The following theorem is one of the first results related to this question. One of its classical proofs relies on some properties of the continued fraction expansion [52, Chap. 6, Theorem 6.24].

Theorem 3.1. *For all $x \in \mathbb{R} - \mathbb{Q}$, there is a sequence $(p_n/q_n)_{n \geqslant 1}$ of rational numbers satisfying*

$$|x - p_n/q_n| \leqslant 1/(2q_n^2) \quad and \quad \lim_{n \to +\infty} |q_n| = +\infty.$$

Can one find a function $\psi(n)$ converging to zero faster than $1/n^2$? The answer is "No" as shown in the following exercise:

Exercise 3.2. Prove that for all $p \in \mathbb{Z}$ and q in $\mathbb{N}^*$, the following inequality is satisfied:

$$|\sqrt{2} - p/q| \geqslant 1/(4q^2).$$

The function ψ that we are looking for, therefore satisfies

$$1/4 \leqslant n^2 \psi(n) \leqslant 1/2.$$

This naturally leads us to define for each irrational number x the quantity

$$\nu(x) = \inf\Big\{ c > 0 \mid \exists\, (p_n/q_n)_{n \geqslant 1} \in \mathbb{Q},$$

$$|x - p_n/q_n| \leqslant c/q_n^2 \text{ and } \lim_{n \to +\infty} |q_n| = +\infty \Big\}.$$

By Theorem 3.1, this quantity is less than $1/2$ for all x. The following theorem is more precise. It can be proved, for example by associating a sequence of circles to the sequence of rational numbers given by the continued fraction expansion, and by studying their relative positions [52, Chap. 6, Theorem 6.25] (see also [30]).

Theorem 3.3. *For every irrational number x, one has*

$$\nu(x) \leqslant 1/\sqrt{5}.$$

Also $\nu(x) = 1/\sqrt{5}$ if and only if there exist a, b, c, d in $\mathbb{Z}$ such that

$$ac - bd = 1 \quad and \quad x = \frac{a\mathcal{N} + b}{c\mathcal{N} + d}, \quad where \; \mathcal{N} = (1 + \sqrt{5})/2 \; is \; the \; golden \; ratio.$$

Among the rational numbers, the *badly approximated real numbers* x for which $\nu(x)$ is strictly positive are of special interest. For example, this is the case for $\sqrt{2}$.

The following theorem relates this property to a property of the sequence of integers $(n_i)_{i \geqslant 0}$ associated with the continuous fraction expansion of x. A proof of this theorem is given, for example, in [24, Theorem 2.20]. Our proof is not very different from the cited example.

Theorem 3.4. *Let x be an irrational number. The following are equivalent:*

(i) *the sequence $(n_i)_{i \geqslant 0}$ is bounded;*
(ii) *the real number x is badly approximated.*

3.2 Hyperbolic proofs of Theorems 3.1, 3.3 and 3.4

The proofs of Theorems 3.1 and 3.3 that we propose are not more elementary than the originals! Our purpose here is not to gain simplicity but to illustrate the fact that the mathematical world is not compartmentalized.

In the rest of this section, $\Gamma = \mathrm{PSL}(2, \mathbb{Z})$ and $p(z) = z+1$. This translation generates the stabilizer Γ_∞ of the point ∞ in Γ.

Recall (from Lemma I.3.19) that the Euclidean diameter of the image by an isometry $\gamma \in \Gamma - \Gamma_\infty$ of the horocycle H_t centered at the point $\gamma(\infty)$ is $1/(c^2(\gamma)t)$, where $c(\gamma)$ is equal to the absolute value of the coefficient c in $\gamma(z)$ written in the form $\gamma(z) = (az + b)/(cz + d)$, with $ad - bc = 1$.

Let $x \in \mathbb{R}$, we have (Fig. VII.12):

$$(*) \qquad (\infty x) \cap \gamma(H_t) \neq \varnothing \Longrightarrow |x - a/c(\gamma)| \leqslant 1/(2tc^2(\gamma)).$$

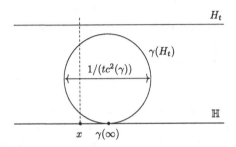

Fig. VII.12.

The following lemma provides a gateway between approximation theory and the study of geodesic rays on the surface $S = \Gamma\backslash\mathbb{H}$. Recall that π is the projection from $\mathbb{H}$ to S.

Lemma 3.5. *Let $x \in \mathbb{R}$ and let $(r(s))_{s\in\mathbb{R}}$ be an arclength parametrization of the oriented geodesic (∞x). The following are equivalent:*

(i) *there exists a sequence $(s_n)_{n\geqslant 1}$ of positive real numbers such that $\lim_{n\to+\infty} s_n = +\infty$ and $\pi(r(s_n))$ belongs to the horocycle $\pi(H_t)$;*
(ii) *there exists a sequence $(\gamma_n)_{n\geqslant 1}$ in $\Gamma - \Gamma_\infty$ such that*

$$|x - \gamma_n(\infty)| \leqslant 1/(2tc^2(\gamma_n)) \quad and \quad \lim_{n\to+\infty} c(\gamma_n) = +\infty.$$

Proof.

(i) $\Rightarrow$ (ii). We will essentially recycle the arguments used in the proof of Proposition I.3.20. The fact that $\pi(r(s_n))$ belongs to $\pi(H_t)$ means that there exists $\gamma_n \in \Gamma$ such that

$$r(s_n) \in \gamma_n(H_t).$$

The geodesic (∞x) intersects $\gamma_n(H_t)$, hence $|x - \gamma_n(\infty)| \leqslant 1/(2tc^2(\gamma_n))$. Since s_n goes to $+\infty$ and $\gamma_n(H_t)$ does not meet H_t, then the sequence $(c(\gamma_n))_{n \geqslant 1}$ is not bounded.

(ii) $\Rightarrow$ (i). By hypothesis, the geodesic (∞x) intersects each circle $\gamma_n(H_t)$. If we let s_n denote the largest real number s such that $r(s) \in \gamma_n(H_t)$, then $r(s_n) = x + ibe^{-s_n}$, where b is a fixed real number > 0. Let R_n denote the Euclidean ray of $\gamma_n(H_t)$. The center of this Euclidean circle is the point $\gamma_n(\infty) + iR_n$, thus the following equation holds:

$$(x - \gamma_n(\infty))^2 + (be^{-s_n} - R_n)^2 = R_n^2.$$

The sequences $(x - \gamma_n(\infty))_{n \geqslant 1}$ and $(R_n)_{n \geqslant 1}$ converge to 0. It follows that $\lim_{n \to +\infty} s_n = +\infty$.   □

We are now ready to prove Theorems 3.1 and 3.3.

Proof (of Theorem 3.1). Let $x \in \mathbb{R} - \mathbb{Q}$. We will show that there exists a sequence $(s_n)_{n \geqslant 1}$ of positive real numbers such that

$$\pi(r(s_n)) \in \pi(H_1) \quad \text{and} \quad \lim_{n \to +\infty} s_n = +\infty.$$

We argue by contradiction. Suppose that there exists $z \in (\infty x)$ such that $\pi([z, x))$ does not intersect $\pi(H_1)$. Lift this situation up to $\mathbb{H}$ and consider the elliptic isometry r in Γ, of order 3, defined by $r(z) = (z - 1)/z$. The isometries r and r^2 map H_1 to circles of Euclidean diameter 1, tangent to the points 1 and 0 respectively, and tangent to each other at the point $1/2 + i/2$ (Fig. VII.13).

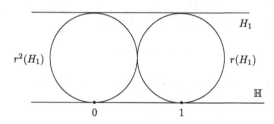

Fig. VII.13.

As shown in Chap. II (Exercise II.3.8), the set Δ defined by

$$\Delta = \{z \in \mathbb{H} \mid 0 \leqslant \operatorname{Re} z \leqslant 1, |z| \geqslant 1 \text{ and } |z - 1| \geqslant 1\},$$

is a fundamental domain of Γ (Fig. VII.14). It follows that there exists $\gamma \in \Gamma$ such that $\gamma(z)$ belongs to Δ. The point x does not belong to the Γ-orbit of the point ∞ and the ray $[\gamma(z), \gamma(x))$, which is a circular arc, does not intersect H_1. Hence this ray is in $\mathbb{H} - H_1^+$. For the same reasons, it is also in $\mathbb{H} - r(H_1^+)$

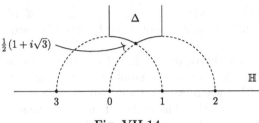

Fig. VII.14.

and $\mathbb{H} - r^2(H_1^+)$. Since $\gamma(z)$ belongs to Δ, this ray is in a compact set, which is impossible.

To achieve the proof, it suffices to apply statement $(*)$, Lemma 3.5 and to note that if γ does not belong to Γ_∞, then $\gamma(\infty)$ is rational of the form $a/c(\gamma)$, where a and $c(\gamma)$ are relatively prime. $\qquad\square$

Proof (of Theorem 3.3). Recall that $\mathbb{R} - \mathbb{Q}$ is the set of conical points of Γ (Property II.3.8). By Lemma 3.5, the quantities $\nu(x)$ and the height $h(x)$ of x (Definition 1.6) are related by the following relation:

$$\nu(x) = 1/(2h(x)).$$

Theorem 3.3 can therefore be restated in terms of the height of geodesic rays in the form

$$\inf_{x \in \mathbb{R} - \mathbb{Q}} h(x) = \sqrt{5}/2.$$

Let us prove this statement. Consider the case where x is the fixed point of a hyperbolic isometry γ in Γ. Choose z on the axis of this isometry, with the condition that $\pi([z, x)) = \pi((\gamma^-\gamma^+))$, where $\pi((\gamma^-\gamma^+))$ is a compact geodesic on S. Thus $h(x)$ is the largest $t > 0$ such that

$$\pi((\gamma^-\gamma^+)) \cap \pi(H_t) \neq \varnothing.$$

In other words, on $\mathbb{H}$ one has the expression (Fig. VII.15)

$$h(x) = \max_{g \in \Gamma} |g(\gamma^-) - g(\gamma^+)|/2.$$

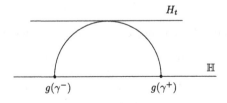

Fig. VII.15.

We write $\gamma(z) = (az + b)/(cz + d)$ with $ad - bc = 1$. Since γ is hyperbolic, we have $c \neq 0$. The following expression is obtained from a simple calculation:

$$(**) \qquad |g(\gamma^-) - g(\gamma^+)| = \sqrt{(a+d)^2 - 4}/c(g\gamma g^{-1}).$$

At the end of Sect. II.4 (Corollary 4.14), we introduced the particular case where x is the golden ratio $\mathcal{N} = (1 + \sqrt{5})/2$. As we saw in that section, this point is fixed by the hyperbolic isometry $\gamma = T_1 T_{-1}$ in Γ, which can also be written in the form

$$\gamma(z) = (2z + 1)/(z + 1).$$

The following equality can be deduced from expression $(**)$:

$$h(\mathcal{N}) = (\sqrt{5}/2) \min_{g \in \Gamma} c(g\gamma g^{-1}).$$

Since for all $g \in \Gamma$ the isometry $g\gamma g^{-1}$ is hyperbolic, we have $c(g\gamma g^{-1}) \neq 0$. Furthermore, this quantity is an integer thus $c(g\gamma g^{-1}) \geqslant 1$. In the particular case where $\gamma = T_1 T_{-1}$, one has $c(\gamma) = 1$. It follows that

$$h(\mathcal{N}) = \sqrt{5}/2,$$

and that for all $x \in \mathbb{R} - \mathbb{Q}$

$$h(x) \leqslant \sqrt{5}/2.$$

We prove now the reverse inequality. Let $x \in \mathbb{R} - \mathbb{Q}$. The definition of the height $h(x)$ implies

$$h(\gamma x) = h(x),$$

for all $\gamma \in \Gamma$. Thus without loss of generality one can assume that x is in $[0, 1]$. Denoted by $([0; n_1, \ldots, n_k])_{k \geqslant 1}$ the continued fraction expansion of x. As in Sect. II.4, we will use the isometries $T_1(z) = z + 1$, $T_{-1}(z) = z/(z + 1)$ and $s(z) = -1/z$. These three isometries are interrelated by the equality $sT_{-1}s = T_1^{-1}$.

Let $k \geqslant 2$ and set $\gamma_k = T_{-1}^{n_1} \cdots T_{(-1)^k}^{n_k}$. Recall the following facts from Exercise II.4.5 and Proposition II.4.6:

- if k is even, then $\gamma_k(0) = [0; n_1, \ldots, n_k]$ and $\gamma_k(\infty) = [0; n_1, \ldots, n_{k-1}]$;
- if k is odd, then $\gamma_k(0) = [0; n_1, \ldots, n_{k-1}]$ and $\gamma_k(\infty) = [0; n_1, \ldots, n_k]$;
- $\lim_{k \to +\infty}[0; n_1, \ldots, n_k] = x$.

Our proof requires the following lemma which relates the integers $(n_i)_{i \geqslant 1}$ to the height $h(x)$.

Lemma 3.6. *There exists a sequence $(g_i)_{i \geqslant 2}$ in Γ such that*

$$g_i^{-1}(\infty x) \cap H_{(n_i+1/(n_{i+1}+1))/2} \neq \varnothing \quad and \quad \lim_{i \to +\infty} c(g_i^{-1}) = +\infty.$$

Proof. If i is even, define $g_i = \gamma_{i-1}$. One has

$$g_i^{-1} = s \circ T_1^{n_{i-1}} \circ T_{-1}^{n_{i-2}} \circ \cdots \circ T_1^{n_1} \circ s.$$

Thus $g_i^{-1}(\infty)$ is the rational $-[0; n_{i-1}, n_{i-2}, \ldots, n_1]$, and the continued fraction expansion of $g_i^{-1}(x)$ is $([n_i; n_{i+1}, \ldots, n_k])_{k \geqslant 1}$.
   If i is odd, define $g_i = \gamma_{i-1} \circ s$. One has:

$$g_i^{-1} = T_{-1}^{n_{i-1}} \circ T_1^{n_{i-2}} \circ \cdots \circ T_1^{n_1} \circ s.$$

Thus $g_i^{-1}(\infty)$ is the rational $[0; n_{i-1}, n_{i-2}, \ldots, n_1]$ and $g_i^{-1}(x)$ has the sequence $(-[n_i; n_{i+1}, \ldots, n_k])_{k \geqslant 1}$ as its continued fraction expansion.
   In these two cases one has:

$$|g_i^{-1}(\infty) - g_i^{-1}(x)| \geqslant n_i + 1/(n_{i+1} + 1),$$

which proves the first part of the lemma.
   The second part of this result arises from the fact that the sequences $(c(g_{2i}))_{i \geqslant 1}$ and $(c(g_{2i+1}))_{i \geqslant 1}$ are sequences of positive integers which are strictly increasing. □

   Now we return to the proof of our inequality. Consider an arclength parametrization $(r(s))_{s \in \mathbb{R}}$ of the oriented geodesic (∞x). If the sequence $(n_i)_{i \geqslant 1}$ contains infinitely many terms $\geqslant 3$, by Lemmas 3.5 and 3.6, there exists an unbounded sequence $(s_n)_{n \geqslant 1}$ in $\mathbb{R}^+$ such that

$$\pi(r(s_n)) \in \pi(H_{3/2}).$$

Thus $h(x) \geqslant 3/2$ which implies $h(x) > \sqrt{5}/2$.
   Otherwise, after replacing x with $\gamma_i^{-1}(x)$ with $i \geqslant 1$, two cases arise:

- either the sequence $(n_i)_{i \geqslant 1}$ contains infinitely many terms equal to 2 and $1 \leqslant n_i \leqslant 2$ for all $i \geqslant 1$,
- or every term of the sequence $(n_i)_{i \geqslant 1}$ is equal to 1.

   In the first case, by Lemmas 3.5 and 3.6, there exists an unbounded sequence $(s_n)_{n \geqslant 1}$ in $\mathbb{R}^+$ such that

$$\pi(r(s_n)) \in \pi(H_{7/6}).$$

Thus $h(x) \geqslant 7/6$, and in particular, $h(x) > \sqrt{5}/2$.
   In the second case, x is related to the golden ratio. More precisely, one has $T_1(x) = \mathcal{N}$, thus $h(x) = h(\mathcal{N})$ and, by the first part of the proof, $h(x) = \sqrt{5}/2$. □

   Notice that, as we saw in Corollary II.4.14, the projection to S of the axis of the hyperbolic isometry $T_1 T_{-1}$ is the shortest compact geodesic in S. It follows from the proof of Theorem 3.3 that it is also the geodesic that achieves the least height in the cusp $\pi(H_1^+)$.
   It now remains only to prove Theorem 3.4.

Proof (of Theorem 3.4). Let x be an irrational number. We know that $\nu(x)$ and $h(x)$ are related by the following equation:

$$\nu(x) = 1/(2h(x)).$$

Thus x is badly approximately if and only if x is geometrically badly approximated. Recall that, by Corollary 2.2, this property is equivalent to the fact that the ray $\pi([z, x))$ is bounded.

Not (i) $\Rightarrow$ not (ii). Suppose that the sequence $(n_i)_{i \geqslant 1}$ is not bounded. For all $t > 0$, there exists a subsequence $(n_{i_k})_{k \geqslant 1}$ whose terms are all $\geqslant 2t$. Let $(r(s))_{s \in \mathbb{R}}$ be an arclength parametrization of the oriented geodesic (∞x). By Lemmas 3.5 and 3.6, there exists a sequence $(s_n)_{n \geqslant 1}$ in $\mathbb{R}^+$ converging to $+\infty$ such that

$$\pi(r(s_n)) \in \pi(H_t).$$

One can conclude that $h(x)$ is greater than t for all $t > 0$ and hence that x is not geometrically badly approximated.

Not (ii) $\Rightarrow$ not (i). Suppose that $\pi([i, x))$ is not bounded. For all integers $k \geqslant 2$ the surface S minus the cusp $\pi(H_k^+)$ is bounded. Hence there exists $K \geqslant 2$ such that for all $k \geqslant K$ (Fig. VII.16)

$$\pi([i, x)) \cap \pi(H_k) \neq \varnothing.$$

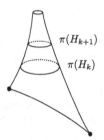

$\pi(H_{k+1})$

$\pi(H_k)$

Fig. VII.16.

On $\mathbb{H}$, this property translates to the existence of a sequence $(g_k)_{k \geqslant K}$ in Γ satisfying

$$g_k^{-1}([i, x)) \cap H_k \neq \varnothing.$$

Since the coefficients of the Möbius transformation g_k are integers, we have $\operatorname{Im} g_k^{-1}(i) \leqslant 1$. Therefore, the ray $g_k^{-1}([i, x))$ crosses H_k at two points, which implies the inequality

$$|\operatorname{Re}(g_k^{-1}(i)) - g_k^{-1}(x)| > k.$$

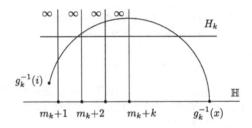

Fig. VII.17.

It follows that this ray intersects at least k vertical geodesics of the form (Fig. VII.17)

$$((m_k + 1)\infty), ((m_k + 2)\infty), \ldots, ((m_k + k)\infty).$$

Hence the ray $[i, x)$ intersects k consecutive Farey lines

$$g_k T_1^{m_k+1}(0\infty), \quad \ldots, \quad g_k T_1^{m_k+k}(0\infty).$$

Returning to the geometric construction of the continued fraction expansion discussed in Sect. II.3, one obtains that for all $k \geqslant K$, there exists $n_i \geqslant k$. □

4 Comments

Artin [3] was one of the first mathematicians who developed this geometric approach to numbers. Some years after, the key idea to built a relationship between fractions and circles appeared in an elementary and informative paper of Ford [30].

The geometric approach to numbers, as presented in Sect. VII.3, allows one to rediscover other classical results. For example, one can obtain properties of the Markov spectrum by relating this object to the lengths of simple (without self intersections), compact geodesics in the quotient of $\mathbb{H}$ over the group $\Gamma(2)$, which was introduced in Chap. II [15, 35, 56]. It also allows some questions of number theory to be formulated in terms of dynamics. Consider for example the open question about characterizing badly approximated algebraic numbers x of degree $n \geqslant 3$, which reduces, on the modular surface, to asking if the associated rays $\pi([i, x))$ are bounded.

The S. Patterson's thesis [50] is one of the founding texts which generalize this approach to Fuchsian groups. More generally, this approach can be adapted for geometrically finite Kleinian groups Γ acting on a pinched Hadamard manifold [38]. Thus it widens the field of the theory of Diophantine approximation, replacing $\mathbb{R}$ with $L(\Gamma)$, and $\mathbb{Q}$ with the orbit of a parabolic point of Γ.

The metric theory of approximations can likewise be approached from this geometric angle, and can be generalized by allowing the role of the Lebesgue measure to be played by a Patterson measure [51, 62]. Using this point of view one can obtain a general version of Khintchine's theorem, whose classical statement establishes a link between the nature of the series $\sum_{n \geqslant 1} \Psi(n)$, where Ψ is a strictly positive decreasing function, and the Lebesgue measure of the set of real numbers x approximated by a sequence of rational numbers $(p_n/q_n)_{n \geqslant 1}$ satisfying $|x - p_n/q_n| \leqslant \Psi(q_n)/q_n$.

Due to the initiative of M.S. Raghunathan, this approach has been similarly developed by G. Margulis, S. Dani and many other mathematicians to solve some problems in Diophantine approximation in $\mathbb{R}^n$ ([60, Chap. IV] and [46]). The dynamical system in play in this context is the action of a closed subgroup of $\mathrm{SL}(n, \mathbb{R})$ on the symmetric space $M^n = \mathrm{SL}(n, \mathbb{Z}) \backslash \mathrm{SL}(n, \mathbb{R})$. One good illustration of the effectiveness of this altered point of view is the proof of the Oppenheim conjecture on the non-degenerate, indefinite, irrational quadratic forms on $\mathbb{R}^n$ $(n \geqslant 3)$. This conjecture asserts that if Q is such a form, then for all $\varepsilon > 0$, there exists $v \in \mathbb{Z}^n - \{0\}$ such that $|Q(v)| \leqslant \varepsilon$. Its proof, due to G. Margulis, requires the topological description of orbits of the group $\mathrm{SO}^0(p, q)$ on M^n, where (p, q) is the signature of Q ([5, Appendix by E. Breuillard] and [32, 47, 60, 17]). In the same spirit, G. Margulis also showed that the Hardy-Littlewood conjecture which stated that for every pair (x, y) in $\mathbb{R}^2$, there exists a sequence of integers $(q_n)_{n \geqslant 1}, (p_n)_{n \geqslant 1}, (r_n)_{n \geqslant 1}$, with $q_n > 0$, such that $\lim_{n \to +\infty} q_n^2 |x - p_n/q_n| |y - r_n/q_n| = 0$, is related to the orbits of the group D of diagonal 3×3 matrices of the form $(e^{t_1 + t_2}, e^{-t_1}, e^{-t_2})$ on M^3 [60, Sect. 30]. At the time of writing of this text, this method has not (yet) produced an answer to this conjecture but has enriched the area of dynamical systems with an open question, namely: Are the bounded orbits of D on M^3 compact?

Appendix A

Basic concepts in topological dynamics

This short introduction to abstract topological dynamics is inspired by M. Alongi's and G.S. Nelson's "Recurrence and topology" [1]. This book contains the solutions of the exercises of this appendix. We encourage a reader who wants to know more on this field to read A. Katok's and B. Hasselblatt's "Introduction to the Modern Theory of Dynamical Systems" [40], and W.H. Gottschalk's and G.A. Hedlund's "Topological dynamics" [34].

Let Y be a topological space. By definition, a flow on Y is a map

$$\phi : \mathbb{R} \times Y \longrightarrow Y$$

satisfying the following conditions:

(i) ϕ is continuous;
(ii) $\phi(t,.) : Y \to Y$ is a homeomorphism for each $t \in \mathbb{R}$;
(iii) $\phi(s, \phi(t, y)) = \phi(s + t, y)$, for all y in Y and any real numbers s, t.

For each real number t, we denote by $\phi_t : Y \to Y$ the map defined for all $y \in Y$ by $\phi_t(y) = \phi(t, y)$.

Exercise A.1. Prove that $\phi_0 = \mathrm{Id}$, and that $\phi_t = \phi_t^{-1}$, for each $t \in \mathbb{R}$.

Many examples arise from smooth vector fields f on smooth manifolds, and are determined by a differential equations of the form

$$f(y) = \frac{dy}{dt},$$

where dy/dt denotes the derivative of a function y with respect to a single independent variable.

In most cases, there exists a unique smooth function $\phi : \mathbb{R} \times Y \to Y$ satisfying

$$\frac{d\phi(t, y)}{dt}(0) = f(y),$$

such that

F. Dal'Bo, *Geodesic and Horocyclic Trajectories*, Universitext,
DOI 10.1007/978-0-85729-073-1, © Springer-Verlag London Limited 2011

(i) $\phi(t,.) : Y \to Y$ is a diffeomorphism for each $t \in \mathbb{R}$;

(ii) $\phi(s, \phi(t, y)) = \phi(s + t, y)$, for all y in Y and any real numbers s, t.

Examples A.2.

(i) If $Y = \mathbb{R}^2$ and $f(y)$ is the constant vector field $\overrightarrow{v} \neq \overrightarrow{0}$, then $\phi(t, y) = y + t\overrightarrow{v}$.

(ii) Notice that the flow $\phi(t, y) = y + t\overrightarrow{v}$ induces a flow Φ on the torus $\mathbb{T}^2 = \mathbb{R}^2/\mathbb{Z}^2$ given by

$$\Phi(t, y \bmod \mathbb{Z}^2) = y + t\overrightarrow{v} \bmod \mathbb{Z}^2.$$

More generally, when Y is a compact smooth manifold, classical theorems for ordinary differential equations guarantee the existence (and uniqueness) of a flow associated to a smooth vector field on Y.

Definition A.3. If $\phi : \mathbb{R} \times Y \to Y$ is a flow, then the *trajectory* (respectively the *positive* or *negative semi-trajectory*) from a point y in Y is the set of points $\phi_t(y)$, where t is in $\mathbb{R}$ (respectively $\mathbb{R}^+$ or $\mathbb{R}^-$).

In Example A.2(i), the trajectory from $y \in \mathbb{R}^2$ is the straight line passing through y with direction $\overrightarrow{v}$.

In Example A.2(ii), the trajectories of Φ on $\mathbb{T}^2$, are the projection on $\mathbb{T}^2$ of the trajectories of Example A.2(i).

Proposition A.4. *Let $\phi : \mathbb{R} \times Y \to Y$ be a flow, if two trajectories have a nonempty intersection, then they are equal.*

Exercise A.5. Prove Proposition A.4.

It follows from Proposition A.4 that the family of all trajectories is a partition of the space Y.

Definition A.6. Let $\phi : \mathbb{R} \times Y \to Y$ be a flow. A point y is a *periodic* point if there exits $T > 0$ such that $\phi_T(y) = y$. The *period* of y is the infimum of such T.

A flow ϕ associated to a non-zero constant vector field on $\mathbb{R}^2$ does not admit periodic points. In contrary, if $\overrightarrow{v} \in \mathbb{Q} \times \mathbb{Q} - \{(0,0)\}$, then all points in the torus $\mathbb{T}^2$ are periodic for the flow induced by ϕ.

Exercise A.7. Prove that a flow on $\mathbb{T}^2$ induced by a non-zero constant vector field $\overrightarrow{v}$ on $\mathbb{R}^2$ has periodic point if and only if $\overrightarrow{v} \in \mathbb{Q} \times \mathbb{Q} - \{(0,0)\}$.

Proposition A.8. *Let $\phi : \mathbb{R} \times Y \to Y$ be a flow. If y is a periodic point, then its trajectory is compact.*

Exercise A.9. Prove Proposition A.8.

Exercise A.10. Let Φ be a flow on $\mathbb{T}^2$ induced by a non-zero constant vector field $\vec{v}$. Prove that, if $\vec{v} \notin \mathbb{Q} \times \mathbb{Q} - \{(0,0)\}$, then each trajectory is dense in $\mathbb{T}^2$.

Notice that, if y is a periodic point for a flow ϕ, or if its trajectory is dense, then there exists an unbounded sequence of real numbers $(t_n)_{n \geqslant 0}$ such that $\lim_{t_n \to \infty} \phi_{t_n}(y) = y$.

More generally we introduce the following definition

Definition A.11. Let $\phi : \mathbb{R} \times Y \to Y$ be a flow. A point y is *non-wandering* if for any neighborhood V of y, there exists an unbounded sequence of real numbers $(t_n)_{n \geqslant 0}$ such that

$$\phi_{t_n} V \cap V \neq \varnothing.$$

We denote by $\Omega_\phi(Y)$ the set of non-wandering points of ϕ.

Notice that in examples (i) and (ii) we have: $\Omega_\phi(\mathbb{R}^2) = \varnothing$ and $\Omega_\Phi(\mathbb{T}^2) = \mathbb{T}^2$. In general the situation is more complicated.

Exercise A.12. Let ϕ be the flow on the closed unit disk $\overline{\mathbb{D}} = \{z \in \mathbb{C} \mid |z| \leqslant 1\}$ associated to the vector field defined in polar coordinates (r, θ) by:

$$\frac{dr}{dt} = r(r - 1) \quad \text{and} \quad \frac{d\theta}{dt} = \theta.$$

Prove that the set of periodic points is $\mathbb{S}^1 \cup \{0\}$ and that $\Omega_\phi(\overline{\mathbb{D}}) = \mathbb{S}^1 \cup \{0\}$. (Hint: see [1, Exercises 2.3.8 and 2.5.12].)

Proposition A.13. *Let* $\phi : \mathbb{R} \times Y \to Y$ *be a flow. The non-wandering set* $\Omega_\phi(Y)$ *is a closed set, invariant with respect to the flow.*

Exercise A.14. Prove Proposition A.13.

In Example A.2(i), no trajectory has accumulation points. More generally we define the notion of divergent points:

Definition A.15. Let $\phi : \mathbb{R} \times Y \to Y$ be a flow. A point y is said to be *divergent* (respectively *positively* or *negatively divergent*) if for all unbounded sequences $(t_n)_{n \geqslant 1}$ in $\mathbb{R}$ (respectively $\mathbb{R}^+$ or $\mathbb{R}^-$), the sequence of points $(\phi_{t_n}(y))_{n \geqslant 1}$ diverges.

Notice that the notion of divergent points makes sense only for non-compact manifolds.

Exercise A.16. Prove that a point y is positively divergent (respectively negatively divergent) if and only if for some $T \in \mathbb{R}$ the function from $[T, +\infty)$ (respectively $(-\infty, T]$) into Y, which sends t to $\phi_t(y)$ is a homeomorphism onto its image (i.e., a topological embedding).

There is no general relations between divergent points and wandering points. Exercise A.12 gives an example without divergent points, where $\Omega_\phi(Y) \neq Y$. It is shown in Chap. III that, when Y is the quotient of $T^1\mathbb{H}$ by the modular group $\mathrm{PSL}(2,\mathbb{Z})$ and ϕ is the geodesic flow, then $\Omega_\phi(Y) = Y$ and there are divergent points.

Definition A.17. A set $M \subset Y$ is *minimal* with respect to the flow ϕ if M is a nonempty closed subset in Y such that for each $m \in M$ its trajectory $\phi_\mathbb{R}(m)$ is dense in M.

Equivalently, a nonempty subset in Y is minimal if it does not contain proper nonempty closed subset, invariant with respect to the flow ϕ.

For example, if y is periodic or is positively and negatively divergent, then its trajectory is minimal.

Appendix B

Basic concepts in Riemannian geometry

This appendix outlines some results proved in [31], which are useful in Chaps. I and VI.

Let M be a connected smooth manifold. A *Riemannian metric* on M is a family of scalar products $(g_m)_{m \in M}$ defined on each tangent space $T_m M$ and depending smoothly on m. For example, the Euclidean space $\mathbb{R}^n$ is canonically equipped with a Riemannian structure $(g_m)_{m \in M}$, where g_m is the ambient scalar product. More generally, if M is a submanifold of $\mathbb{R}^n$, then the restriction of g_m to each tangent space $T_m M$ induces a Riemannian metric on M. This is the case for example for the torus $\mathbb{T}^2$ viewed as revolution surface in $\mathbb{R}^3$ induced by the map $\psi : \mathbb{R}^2 \to \mathbb{R}^3$ defined by

$$\psi(\theta, \phi) = ((2 + \cos\theta)\cos\phi, (2 + \cos\theta)\sin\phi, \sin\theta).$$

Given a Riemannian metric $(g_m)_{m \in M}$ on M, we are lead to define a *canonical measure* v_g on M. More precisely, let (U_k, ϕ_k) be a chart and consider the local expression of the metric in this chart

$$\sum_{1 \leqslant i,j \leqslant \text{Dim}(M)} g_{ij}^k dx_i dx_j.$$

The volume of the parallelotope generate by the vectors $\partial/\partial x_i$ is $\sqrt{\det(g_{ij}^k)}$. We define the measure v_g as corresponding to the density which is given in the atlas (U_k, ϕ_k) by

$$\sqrt{\det(g_{ij}^k)}L,$$

where L is he Lebesgue measure on $\mathbb{R}^n$.

By definition, the volume of a subset $B \subset M$ is given by

$$\text{vol}(B) = \int_B v_g,$$

when this integral exists.

F. Dal'Bo, *Geodesic and Horocyclic Trajectories*, Universitext,
DOI 10.1007/978-0-85729-073-1, © Springer-Verlag London Limited 2011

For the torus $\mathbb{T}^2$ viewed as revolution surface in $\mathbb{R}^3$, the *area* of a subset $B = \psi(A) \subset \mathbb{T}^2$ associated with the induced metric is given by

$$\mathrm{vol}(B) = \int_A \sqrt{\left| \frac{\partial d\psi}{d\theta} \wedge \frac{\partial d\psi}{d\phi} \right|} \, d\theta \, d\phi,$$

where $\wedge$ is the vector product in $\mathbb{R}^3$.

The notion of *length* of a piecewise C^1 curve $c : [0, a] \to M$ is also well defined and is given by

$$\mathrm{length}(c) = \int_0^a \sqrt{g_{c(t)}(c'(t), c'(t))} \, dt.$$

This notion does not depend on the choice of a regular parametrization.

Using the notion of length, we define a distance on M associated to the Riemannian metric $(g_m)_{m \in M}$. The following proposition is proved in [31, Proposition 2.91].

Proposition B.1. *Let $d : M \times M \to \mathbb{R}^+$ be the map defined for m and m' as the infimum of the lengths of all piecewise C^1 curves from m to m'. This map is a distance on M, which gives back the topology of M.*

In the Euclidean space, straight lines are length minimizing. The curves which (locally) minimize length in a Riemannian manifold are the *geodesics*. Namely we have [31, Corollary 2.94].

Definition B.2. A curve $c : I \subset \mathbb{R} \to M$, parametrized proportional to arclength, is a geodesic if and only if for any $t \in I$ there exists $\varepsilon > 0$ such that $d(c(t), c(t + \varepsilon)) = \mathrm{length}(c_{|[t, t+\varepsilon]})$.

For the metric on $\mathbb{T}^2$ viewed as revolution surface in $\mathbb{R}^3$ meridian lines and parallels ($\theta = $ constant) parameterized proportional to length are geodesics [31, Exercise 2.83].

A diffeomorphism f between a Riemannian manifold $(M, (g_m)_{m \in M})$ and a smooth manifold M' induces a metric on M' defined for $m' \in M'$ and $\overrightarrow{u}', \overrightarrow{v}' \in T'_m M'$ by

$$g'_{m'}(\overrightarrow{u}', \overrightarrow{v}') = g_{f^{-1}(m')}(T_{m'} f^{-1}(\overrightarrow{u}'), T_{m'} f^{-1}(\overrightarrow{v}')).$$

The Riemannian manifolds $(M, (g_m)_{m \in M})$ and $(M', (g'_{m'})_{m' \in M'})$ are isometric in the following sense

Definition B.3. Let $(M, (g_m)_{m \in M})$ and $(M', (g'_{m'})_{m' \in M'})$ be two Riemannian manifolds. A map $f : M \to M'$ is an *isometry* (resp. *local isometry*) if f is a diffeomorphism (resp. *local diffeomorphism*), satisfying the following relation for any $m \in M$ and $\overrightarrow{u}, \overrightarrow{v} \in T_m M$

$$g'_{f(m)}(T_m f(\overrightarrow{u}), T_m f(\overrightarrow{v})) = g_m(\overrightarrow{u}, \overrightarrow{v}).$$

When $M' = M$, the set of isometries $f : M \to M$ is a group. Let Γ be a discrete group of isometries of M. We suppose that Γ acts on M *freely* (i.e., for $m \in M$ and $g \in \Gamma - \{\mathrm{Id}\}$, $g(m) \neq m$), and *properly* (i.e., for any $m, m' \in M$, if $m' \notin \Gamma m$, then there exist two neighborhoods $V(m)$ and $V(m')$ such that $gV \cap V' = \varnothing$, for any $g \in \Gamma$). Under these conditions, there exists an unique Riemannian metric on $\Gamma \backslash M$ such that the canonical projection of M onto $\Gamma \backslash M$ is a smooth covering map and a local isometry [31, Proposition 2.20]. For example, if Γ is a group of translations associated to a basis of $\mathbb{R}^2$, we obtain a Riemannian metric on the torus $\mathbb{T}^2$, which is said to be *flat* [31, Exercise 2.25]. For a flat Riemannian metric on $\mathbb{T}^2$, the geodesics are the projections of the straight lines of $\mathbb{R}^2$ parameterized proportional to length.

More generally, we have [31, Proposition 2.81].

Proposition B.4. *If Γ is a discrete group of isometries of $(M, (g_m)_{m \in M})$ acting freely and properly on M, then the geodesics of $\Gamma \backslash M$ are the projections of the geodesics of M, and the geodesics of M are the lifting of those of $\Gamma \backslash M$.*

References

[1] Alongi, J. M., Nelson, G. S.: Recurrence and Topology. Graduate Studies in Mathematics, vol. 85. American Mathematical Society, Providence (2007), pp. 221

[2] Arnoux, P.: Le codage du flot géodésique sur la surface modulaire. Enseign. Math. (2) 40(1–2), 29–48 (1994)

[3] Artin, E.: Ein mechanisches System mit quasiergodischen Bahnen. In: Mathematisches Seminar, Hamburg, pp. 171–175 (1924)

[4] Auslander, L., Green, L., Hahn, F.: Flows on Homogeneous Spaces. Annals of Mathematics Studies, vol. 53. Princeton University Press, Princeton (1963), pp. 107

[5] Babillot, M.: Points entiers et groupes discrets: de l'analyse aux systèmes dynamiques. In: Rigidité, Groupe Fondamental et Dynamique. Panoramas & Synthèses, vol. 13, pp. 1–119. Société Mathématique de France, Paris (2002). With an appendix by Emmanuel Breuillard

[6] Ballmann, W., Gromov, M., Schroeder, V.: Manifolds of Nonpositive Curvature. Progress in Math., vol. 61. Birkhäuser, Basel (1985)

[7] Beardon, A.F.: The Geometry of Discrete Groups. Graduate Texts in Math., vol. 91. Springer, New York (1995), pp. 337

[8] Bedford, T., Keane, M., Series, C. (eds.): Ergodic Theory, Symbolic Dynamics, and Hyperbolic Spaces. Oxford Science, New York (1991). The Clarendon Press/Oxford University Press (1991). Papers from the Workshop on Hyperbolic Geometry and Ergodic Theory held in Trieste, April 17–28, 1989

[9] Berger, M., Gostiaux, B.: Géométrie Différentielle: Variétés, Courbes et Surfaces, 2nd edn. Mathématiques. Presses Universitaires de France, Paris (1992), pp. 513

[10] Bonahon, F., Otal, J.-P.: Variétés hyperboliques à géodésiques arbitrairement courtes. Bull. Lond. Math. Soc. 20(3), 255–261 (1988)

[11] Bourdon, M.: Structure conforme au bord et flot géodésique d'un CAT(-1)-espace. Enseign. Math. (2) 41(1–2), 63–102 (1995)

F. Dal'Bo, *Geodesic and Horocyclic Trajectories*, Universitext,
DOI 10.1007/978-0-85729-073-1, © Springer-Verlag London Limited 2011

[12] Bowditch, B. H.: Geometrical finiteness with variable negative curvature. Duke Math. J. 77(1), 229–274 (1995)

[13] Brouzet, R., Boualem, H.: La Planète R—Voyage au Pays des Nombres Réels. Dunod, Paris (2002)

[14] do Carmo, M.P.: Riemannian Geometry. Mathematics: Theory & Applications. Birkhäuser, Boston (1992), pp. 300

[15] Cassels, J. W. S.: An Introduction to Diophantine Approximation. Cambridge Tracts in Mathematics and Mathematical Physics, vol. 45. Cambridge University Press, New York (1957), pp. 166

[16] Conze, J.-P., Guivarc'h, Y.: Densité d'orbites d'actions de groupes linéaires et propriétés d'équidistribution de marches aléatoires. In: Rigidity in Dynamics and Geometry, Cambridge, 2000, pp. 39–76. Springer, Berlin (2002)

[17] Courtois, G., Dal'Bo, F., Paulin, F.: Sur la dynamique des groupes de matrices et applications arithmétiques. In: Berline, N., Plagne, A., Sabbah, C. (eds.) Journées X-UPS. Éditions de l'École Polytechnique, Palaiseau (2007)

[18] Dal'Bo, F.: Remarques sur le spectre des longueurs d'une surface et comptages. Bol. Soc. Bras. Mat. (NS) 30(2), 199–221 (1999)

[19] Dal'Bo, F.: Topologie du feuilletage fortement stable. Ann. Inst. Fourier (Grenoble) 50(3), 981–993 (2000)

[20] Dal'Bo, F., Peigné, M.: Comportement asymptotique du nombre de géodésiques fermées sur la surface modulaire en courbure non constante. In: Méthodes des Opérateurs de Transfert: Transformations Dilatantes de l'Intervalle et Dénombrement de Géodésiques Fermées. Astérisque, vol. 238, pp. 111–177. Société Mathématique de France, Paris (1996)

[21] Dal'Bo, F., Peigné, M.: Some negatively curved manifolds with cusps, mixing and counting. J. Reine Angew. Math. 497, 141–169 (1998)

[22] Dal'Bo, F., Starkov, A.N.: On noncompact minimal sets of the geodesic flow. J. Dyn. Control Syst. 8(1), 47–64 (2002)

[23] Dal'Bo, F., Otal, J.-P., Peigné, M.: Séries de Poincaré des groupes géométriquement finis. Israel J. Math. 118, 109–124 (2000)

[24] Dani, S. G.: Divergent trajectories of flows on homogeneous spaces and Diophantine approximation. J. Reine Angew. Math. 359, 55–89 (1985)

[25] Dieudonné, J. (ed.): Abrégé D'histoire des Mathématiques 1700–1900, 2nd edn. Hermann, Paris (1986). pp. 517

[26] Eberlein, P.: Geodesic flows on negatively curved manifolds. I. Ann. Math. (2) 95, 492–510 (1972)

[27] Eberlein, P.: Geodesic flows on negatively curved manifolds. II. Trans. Am. Math. Soc. 178, 57–82 (1973)

[28] Eberlein, P., O'Neill, B.: Visibility manifolds. Pac. J. Math. 46, 45–109 (1973)

[29] Ferte, D.: Flot horosphérique des repères sur les variétés hyperboliques de dimension 3 et spectre des groupes kleiniens. Bull. Braz. Math. Soc. (NS) 33(1), 99–123 (2002)

[30] Ford, L. R.: Fractions. Am. Math. Mon. 45(9), 586–601 (1938)

[31] Gallot, S., Hulin, D., Lafontaine, J.: Riemannian Geometry, 3rd edn. Universitext. Springer, Berlin (2004), pp. 322

[32] Ghys, É.: Dynamique des flots unipotents sur les espaces homogènes. In: Séminaire Bourbaki, vol. 1991/92. Astérisque, vol. 206, pp. 93–136. Société Mathématique de France, Paris (1992). Exp. No. 747

[33] Godbillon, C.: Éléments de Topologie Algébrique. Hermann, Paris (1971), pp. 249

[34] Gottschalk, W.H., Hedlund, G.A.: Topological Dynamics. American Mathematical Society Colloquium Publications, vol. 36. American Mathematical Society, Providence (1955), pp. 151

[35] Haas, A.: Diophantine approximation on hyperbolic Riemann surfaces. Acta Math. 156(1–2), 33–82 (1986)

[36] de la Harpe, P.: Free groups in linear groups. Enseign. Math. (2) 29(1–2), 129–144 (1983)

[37] Hedlund, G. A.: Fuchsian groups and transitive horocycles. Duke Math. J. 2(3), 530–542 (1936)

[38] Hersonsky, S., Paulin, F.: Hausdorff dimension of Diophantine geodesics in negatively curved manifolds. J. Reine Angew. Math. 539, 29–43 (2001)

[39] Katok, A., Climenhaga, V.: Lectures on Surfaces. Student Mathematical Library, vol. 46. American Mathematical Society, Providence (2008), pp. 286

[40] Katok, A., Hasselblatt, B.: Introduction to the Modern Theory of Dynamical Systems. Encyclopedia of Mathematics and Its Applications, vol. 54. Cambridge University Press, Cambridge (1995), pp. 802

[41] Katok, S.: Fuchsian Groups. Chicago Lectures in Mathematics. University of Chicago Press, Chicago (1992), pp. 175

[42] Khintchine, A.Ya.: Continued Fractions. P. Noordhoff, Groningen (1963), pp. 101

[43] Kulikov, M.: The horocycle flow without minimal sets. C.R. Math. Acad. Sci. Paris 338(6), 477–480 (2004)

[44] Lalley, S. P.: Renewal theorems in symbolic dynamics, with applications to geodesic flows, non-Euclidean tessellations and their fractal limits. Acta Math. 163(1–2), 1–55 (1989)

[45] Long, D.D., Reid, A.W.: Pseudomodular surfaces. J. Reine Angew. Math. 552, 77–100 (2002)

[46] Margulis, G. A.: Dynamical and ergodic properties of subgroup actions on homogeneous spaces with applications to number theory. In: Proceedings of the International Congress of Mathematicians, Kyoto, 1990, pp. 193–215. Math. Soc. Japan, Tokyo (1991)

[47] Morris, D. W.: Ratner's Theorems on Unipotent Flows. Chicago Lectures in Mathematics. University of Chicago Press, Chicago (2005), pp. 203

[48] Nicholls, P. J.: The Ergodic Theory of Discrete Groups. London Mathematical Society Lecture Note Series, vol. 143. Cambridge University Press, Cambridge (1989), pp. 221

[49] Pansu, P.: Le flot géodésique des variétés riemanniennes à courbure négative. In: Séminaire Bourbaki, vol. 1990/91. Astérisque, vol. 201–203, pp. 269–298. Société Mathématique de France, Paris (1991). Exp. No. 738

[50] Patterson, S.J.: Diophantine approximation in Fuchsian groups. Philos. Trans. R. Soc. Lond. Ser. A 282(1309), 527–563 (1976)

[51] Patterson, S.J.: The limit set of a Fuchsian group. Acta Math. 136(3–4), 241–273 (1976)

[52] Rademacher, H.: Lectures on Elementary Number Theory. R.E. Krieger, Huntington (1977), pp. 146. Reprint of the 1964 original

[53] Ratner, M.: Raghunathan's conjectures for SL(2, ℝ). Israel J. Math. 80(1–2), 1–31 (1992)

[54] Roblin, T.: Ergodicité et Équidistribution en Courbure Négative. Mém. Soc. Math. Fr. (NS), vol. 95. Société Mathématique de France, Paris (2003), pp. 96

[55] Series, C.: Symbolic dynamics for geodesic flows. Acta Math. 146(1–2), 103–128 (1981)

[56] Series, C.: The geometry of Markoff numbers. Math. Intell. 7(3), 20–29 (1985)

[57] Series, C.: The modular surface and continued fractions. J. Lond. Math. Soc. (2) 31(1), 69–80 (1985)

[58] Sinaĭ, Ja.G.: Markov partitions and U-diffeomorphisms. Funkc. Anal. Prilož. 2(1), 64–89 (1968)

[59] Starkov, A. N.: Fuchsian groups from the dynamical viewpoint. J. Dyn. Control Syst. 1(3), 427–445 (1995)

[60] Starkov, A. N.: Dynamical Systems on Homogeneous Spaces. Translations of Mathematical Monographs, vol. 190. American Mathematical Society, Providence (2000), pp. 243. Translated from the 1999 Russian original by the author

[61] Sullivan, D.: The density at infinity of a discrete group of hyperbolic motions. Publ. Math. Inst. Hautes Études Sci. 50, 171–202 (1979)

[62] Sullivan, D.: Disjoint spheres, approximation by imaginary quadratic numbers, and the logarithm law for geodesics. Acta Math. 149(3–4), 215–237 (1982)

Index

F. Dal'Bo, *Geodesic and Horocyclic Trajectories*, Universitext,
DOI 10.1007/978-0-85729-073-1, © Springer-Verlag London Limited 2011